Innovation and Finance

I0867140

Innovation and finance are in a symbiotic and twin-track relationship: a well-functioning financial system spurs innovation by identifying and funding stimulating entrepreneurial activities which trigger economic growth. Innovations also open up profitable opportunities for the financial system. These mutual dynamics cause and need innovative adaptations in the financial system in order to better deal with the changing requirements of a knowledge-based economy. This volume comprises different contributions which focus on the central imperative of this evident connection between financial markets and innovation which, despite its importance, is only barely considered in academia, as well as in practice so far.

This book is about the mutual interdependence of innovation processes and finance. This interdependent relationship is characterized by a high degree of complexity which stems, on the one hand, from the truly uncertain character of innovation and, on the other hand, from the different timescales in both domains. Whereas innovation processes are long-term and experimental, financial markets are interested in shortening time horizons in order to optimize financial investments. Economies which do not manage to align the two realms of their economic system are in danger of ending up in either financial bubbles or economic stagnation. The chapters of this book deal with different aspects of this complex interrelationship between innovation and finance, highlighting, for example the role of stock markets, venture capital and international financial transactions, as well as the historical co-development of the financial and industrial domains. Thus far, the communities in economics dealing with both issues are almost completely disconnected.

The book brings together economic research dealing with the interface between innovation and finance and highlights the importance of the Neo-Schumpeterian perspective. This topic is of particular interest in the current economic crisis affecting the Eurozone and its currency. Most of the policy instruments discussed and implemented so far are focused on short-run targets. This discussion of the relationship between innovation and finance suggests a long-run perspective to create new potentials for economic growth and a sustainable way out of the economic crisis.

Andreas Pyka is Chair of Economics of Innovation at the University of Hohenheim, Germany.

Hans-Peter Burghof is Chair of the Banking and Finance Department at the University of Hohenheim, Germany.

Routledge/Lisbon Civic Forum Studies in Innovation

Series Editors:

Horst Hanusch
University of Augsburg, Germany
Tibor Palánkai
Corvinus University of Budapest, Hungary
Ryszard Wilczynski
Warsaw University of Finance and Management, Poland
Andreas Pyka
University of Hohenheim, Germany
and
Carmen Ruiz Viñals
Universitat Abat Oliba, Spain

Social Innovation
New forms of organisation in knowledge-based societies
Edited by Carmen Ruiz Viñals and Carmen Parra Rodríguez

Innovation and Finance
Edited by Andreas Pyka and Hans-Peter Burghof

Innovation and Finance

Edited by Andreas Pyka and
Hans-Peter Burghof

Routledge
Taylor & Francis Group
LONDON AND NEW YORK

First published 2014
by Routledge
2 Park Square, Milton Park, Abingdon, Oxfordshire OX14 4RN

Simultaneously published in the USA and Canada
by Routledge
711 Third Avenue, New York, NY 10017

First issued in paperback 2016

Routledge is an imprint of the Taylor & Francis Group, an informa business

British Library Cataloguing in Publication Data
A catalogue record for this book is available from the British Library

Library of Congress Cataloging in Publication Data
Innovation and finance/edited by Andreas Pyka and Hans-Peter Burghof.
pages cm.
1. Technological innovations–Finance. 2. New business enterprises–Finance. 3. Venture capital. 4. New products–Finance. I. Pyka, Andreas. II. Burghof, Hans-Peter.
HC79.T4I54576 2013
338.5–dc23 2013003618

ISBN 13: 978-1-138-23132-0 (pbk)
ISBN 13: 978-0-415-69685-2 (hbk)

Typeset in Times New Roman
by Wearset Ltd, Boldon, Tyne and Wear

Contents

Illustrations

Figures

Tables

Contributors

Hans-Peter Burghof, Professor of Banking and Finance at the University of Hohenheim, Germany.

Pelin Demirel, Lecturer in Industrial Economics at University of Nottingham, UK.

Gunnar Eliasson, Professor Emeritus in Industrial Economics/Dynamics at the Royal Institute of Technology (KTH) in Stockholm, Sweden.

Mariana Mazzucato, R.M. Phillips Professor in Science and Technology Policy (SPRU Department) at the University of Sussex, UK.

Jorge Niosi, Professor, Department of Management and Technology, Université du Québec à Montréal, Canada Research Chair on the Management of Technology.

Carlota Perez, Centennial Professor, London School of Economics, UK, and Professor of Technology and Development, Technological University of Tallinn, Estonia.

Andreas Pyka, Professor in Innovation Economics at the University of Hohenheim, Germany.

Valérie Revest, Associate Professor in Finance and Economics, University of Lyon 2, France.

Alessandro Sapio, Assistant Professor of Economics at the Parthenope University of Naples, Italy.

Pier Paolo Saviotti, Guest Professor, Department of Economics, University of Hohenheim, Germany; Guest Professor, ECIS, Technological University, Eindhoven, Germany; Associate Research Fellow, INRA_GAEL Grenoble, France and GREDEG-CNRS, Sophia Antipolis, France.

Morris Teubal, Professor at the Department of Economics, Hebrew University, Jerusalem, Israel.

Ryszard Wilczynski, Professor in Economics at the Warsaw University of Finance and Management, Poland.

Introduction[1]

Andreas Pyka and Hans-Peter Burghof

The first two decades of the twenty-first century are characterized by ongoing critical developments in the financial realms of economies, and a continuous financial crisis which strongly affects the highly developed countries. Contrary to the previous financial crises of the final 30 years of the twentieth century, which had their origins in less developed regions and stemmed from an imbalanced integration within the world economy, the recent crisis originated in the core of the capitalistic organized world. The European Union is one of the centres of these developments.

The cutting loose of public debts in several European economies has become a major originator and intensifier of the crisis, potentially threatening the future of the common currency. Obviously, a solution has to include the re-organization of global financial transactions and the implementation of strict rules concerning public debt and new indebtedness. These instruments might help to mitigate the crisis, at least in the short run. In order to reach a sustainable solution, we will most likely have to change the institutional set-up of the European Union in order to create a solid basis for economic development which is in accord with the European welfare states.

One deep reason for the current critical developments is the mediocre innovative performance of some European economies, which is accompanied by a decreasing international competitiveness. In a global environment with rapidly emerging actors such as the BRIC economies, slow development (or even a decrease) is relatively equivalent to falling behind in competitiveness terms. Simultaneously, growing public tasks and inefficiencies in the public sector will imperatively lead to the collapse of the economy.

Given these obvious interrelations, it is even more surprising that, so far, the relationship between innovation and finance has not had a prominent place on the agenda of economics research. Without a severe and sustainable improvement of the innovative performance, the fundamentals for the sound future economic development of Europe become very weak. A conditio-sine-qua-non for stimulating innovative processes, however, is the provision of a financial environment which supports innovative entrepreneurial activities and is capable of dealing with the strong uncertainties of innovation.

Of course, innovation and entrepreneurial activities are not restricted to the industrial pillars of the economies, but also encompass their financial and public pillars (Hanusch and Pyka 2007a). Public sector innovation, public entrepreneurship, social innovation, and innovation and venture financing are some keywords which express the comprehensive nature of the required approach to design the future-orientation of economic systems. This important issue is dealt with in another volume of the Lisbon Civic Forum series (Riuz and Parra 2013).

This volume addresses issues relating to the financial pillar as well as its interfaces with the industrial and public pillars of an economy. Basically, we try to give various answers to the following question: What characterizes the symbiotic and twin-track relationship between innovation and finance? On the one hand, a well-functioning financial system spurs innovation by identifying and funding stimulating entrepreneurial activities which trigger economic growth. On the other hand, innovations open up profitable opportunities for the financial system. These mutual dynamics cause and require innovative adaptations not only in the technological realm, but also in the financial system in order to better deal with the changing requirements of a knowledge-based economy. These mutual relationships are also the root of strong co-evolutionary dynamics between the three pillars of an economy, which encompass positive feedbacks on the micro-, meso- and macro-levels of an economy and are a rich source of complexity. Consequently, no simple answer can be expected, but rather a strong sensitivity towards initial conditions, unexpected delays and unexpected pattern changes in economic development.

The issue of *innovation and finance* has to be considered to be a strong research gap. The relationship and mutual dependence between financial markets and innovation – which, in Schumpeter's (1911) theory of economic development, is crucial for the understanding of economic dynamics – was barely in the focus of economists throughout the twentieth century. Both the literatures on finance and on innovation have developed prolifically over the last 50 years, yet the interfaces between these two pillars of the economic system remain almost unexplored. One of the reasons for this is the different time horizon in the two literatures: whereas in Neo-Schumpeterian (evolutionary) innovation economics, long-time horizons are essential to grasp the structural changing power of technological development, the finance literature remained stuck in the neoclassical paradigm, with its extreme short-term orientation.

Without doubt, the closing of this research gap will allow for creative impulses to emerge from the scientific and political discussions concerning the future-orientation of economies. This research gap was our major inspiration for organizing a symposium of the Lisbon Civic Forum in Stuttgart on 21 October 2010 at the University of Hohenheim. Distinguished scholars from all over the world were invited to connect the two realms of innovation and finance and to develop innovative approaches suited to the analysis of the co-evolutionary relationship between innovation and finance.

The Lisbon Civic Forum[2]

The Lisbon Civic Forum eV (LCF) for Growth and Jobs was founded in Lisbon in February 2005 by a small group of distinguished scholars and professionals from both old and new EU member states. The LCF aims to provide a platform for the exchange of ideas and opinions between academics, policymakers, diplomats, journalists and practitioners. It also serves as a forum for an intense dialogue with civil societies in Europe, spurring knowledge transfer from academia to policy and society.

The activities of the Forum concentrate on the social and economic progress of the European Union in terms of creating jobs, fostering economic growth and best practices in government, society and business, and dissolving economic crises. These objectives comprise the central elements of the Lisbon Strategy, in which the European Union has expressed its great ambition "to become the most competitive and dynamic knowledge-based economic region in the world". The Lisbon Strategy and the multiple, interdependent conditions and hidden impediments to its realization are central topics of the Lisbon Civic Forum agenda.

From the very beginning, the Lisbon Civic Forum (LCF) has followed a comprehensive and interdisciplinary approach: new ways of thinking, innovative and problem-solving approaches to cooperation, and a higher level of social creativity and solidarity are needed to cope with the main social, economic and institutional challenges that Europe faces in the context of globalization and the transformation of societies and economies. Such approaches are developed by the interdisciplinary and international expert groups of the LCF. The LCF also extends to and include experts from candidate countries as well as from other countries with an EU perspective.

Europe is both the agent and the subject of unprecedented rapid and fundamental social, economic, environmental and institutional transformation. This formidable and robust change undermines conventional wisdom and knowledge, endangers economic development, and jeopardizes job security. At the same time, it provides unprecedented opportunities for larger chains of cross-border cooperation, social innovation involving new players to meet the new challenges, and prolific conditions for a peaceful and economically viable future for the continent.

For the Lisbon Civic Forum, the main strengths of the EU lie in:

1 the single, internal market across the Union and the common European institutions, including the common currency in the Euro countries. The transition towards knowledge-based societies, which advances as rapidly as the process of globalization, requires large investments in education, research and development activities. Neither national firms of normal size nor smaller countries will be able to make such large investments. Only the larger regions will be able to successfully make such great changes. The argument of size, therefore, especially with regard to the number of customers in an internal market, will affect the capabilities of regions to compete in global markets. Only recently, and as a result of its various enlargements, has the EU emerged with the necessary population size and the necessary production and market capacities to compete with other globally oriented economic regions at the international level. Due to its new

size, the EU offers greater opportunities to businesses in its member countries to act and succeed as global and responsible players on international markets.

2 the diversity of economic structures at national and regional levels. The existence of diversity is an essential precondition for growth and for the augmentation of welfare and job creation in capitalist economies. This applies to firms, nations and large economic regions. Diversity builds the basis for the successful division of labour and for the introduction of innovations in society and the economy. Nurturing diversity, therefore, is fundamental to change and progress in all fields of human life. The distinctive regional diversity of the EU is one of its main advantages, offering the best opportunities for the creation of regional clusters as the driving force behind economic development.

3 intensive cooperation and solidary responsibility, which offers new social and ecological responses to globalization. Since the beginning of the process of European integration, economic cooperation has been remarkably successful. Nations, regions and civil societies have developed a stable and peaceful culture of cooperation. This new culture of complex cooperation and interdependence constitutes a major strength of the EU and has to be continuously expanded and enriched.

Comprehensive Neo-Schumpeterian economics

Since the 1980s, questions of economic growth and economic development have experienced a renaissance in economics, after almost 25 years of silence around this topic. The seemingly unsolvable problems of overcoming decreasing rates of marginal capital productivity and the methodological problems of an aggregate production function (e.g. Sraffa 1960) put economic growth theory offside after Solow's promising start in the 1950s (Solow 1956). In the revived discussion on economic growth the major drivers of quantitative growth, as well as – as some protagonists of Neo-Schumpeterian economics claim – of qualitative growth (e.g. Saviotti 1996), are technological, organizational and institutional innovations. Economists widely agree on this today.

Despite this general agreement on the important role of innovations, two different schools of thought occupy this research program:

1 Neoclassical growth theory experienced a proper rejuvenation with the so-called New Growth Theory (see, among others, Romer 1987; Lucas 1988). New Growth theoretical approaches allow getting rid of the major problems of decreasing marginal capital productivity and the convergence of growth rates by considering positive feedback effects (e.g. the so-called technological spillovers or the explicit consideration of human capital) emerging in innovation processes. Although theoretical inconsistencies cannot be denied – in particular, the concept of positively interpreted spillovers on a macro-economic level cannot be complemented on a micro-economic level, where

technological spillovers generally find a negative interpretation (see Pyka *et al.* 2009) – New Growth Theory is considered to fit well within the theoretical framework of neoclassical theory, which considerably supported its diffusion.

2 The alternative approach in Neo-Schumpeterian economics, which started almost at the same time, has chosen a radically different approach, without trying to integrate into the body of theory of neoclassical economics (e.g. Nelson and Winter 1982). Referring to the Schumpeter's (1911) Theory of Economic Development, the contradictoriness of economic development driven by innovation with concepts such as Olympic rationality and economic equilibria is emphasized. Economic growth driven by innovation is compulsorily accompanied by structural change endogenously caused by the purposeful and sometimes erroneous actions and interactions of economic agents, i.e. knowledge generation and diffusion processes. Increasing efficiency on a sectoral level (i.e. process innovation) raises resources which are used for explorative purposes (i.e. product innovation), which might lead to the emergence of new industries supporting long-run economic growth by simultaneously triggering qualitative development (Saviotti and Pyka 2004). These dynamics are to be observed on a micro-economic level (entrepreneurship) and, in the case of success, manifest themselves on the sectoral level (industry life cycles). However, what is measured on the macroeconomic level as economic growth is only the average of an economy from structural dynamics on the meso-level, and very likely does not tell anything on the causes of development.

Today, Neo-Schumpeterian economics has become an independent and widely recognized research programme (e.g. Dosi *et al.* 1988; Fagerberg *et al.* 2005; Dopfer 2005; Hanusch and Pyka 2007a) which has considerable influence upon the design of innovation and technology policy, particularly by international organizations such as the OECD (OECD 1991), the World Bank (World Bank 1999) and the European Commission. Since the mid-1990s, technology and innovation policy cannot be analysed without Neo-Schumpeterian concepts such as technological clusters (e.g. Braunerhjelm and Feldman 2007), innovation networks (e.g. Pyka 2002), and entrepreneurship (e.g. Grebel *et al.* 2003).

Until recently, however, the innovation-orientation of the Neo-Schumpeterian approach was applied almost exclusively to manufacturing and service industries. In Hanusch and Pyka (2007b) we show that the innovation-orientation of the industrial sector is only one prerequisite for economic growth and development. The growth success of an economy similarly depends upon the innovation-orientation or respective future-orientation of financial markets as well as upon the public sector. Economic growth and development are carried by these three pillars of economic systems, which are encompassed by the bracket of true uncertainty (Knight 1921), which is inseparably connected to all kinds of innovative development processes. The intrinsic uncertainty of innovation processes is the major reason why, when it comes to true innovation, mainstream economic approaches cannot be applied either for analysing industries or financial markets or for the activities of the public sector. The concept of rationality as applied in neoclassical economics is not applicable in uncertain situations. The Olympic rationality of neoclassical economics leads to a *pathological pessimism* (Erdmann 1993) concerning any kind of innovation. Without a general willingness to innovate (i.e. a willingness to deal with the *ex-ante* non-predictable possibility of failure and economic losses), any innovative behaviour becomes impossible.

From this, one can easily see that for an economic analysis of the potentials for growth and development of economies one cannot apply the idea of innovation to industrial sectors only. The innovation-orientation has to be transferred to the financial markets and the public sector, as well as to the important mutual influences between these three realms of economic development. With this transfer of innovation – respectively, future-orientation and the accompanying uncertainty – we can roughly outline the scope of the contributions to this volume.

The contributions

In her chapter – "Financial bubbles, crises, and the role of government in unleashing golden ages" – Carlota Perez gives a precise classification of the current global economic crisis in long-run economic development. She argues that the recent financial crisis is not an ordinary one, but a once-in-a-half-century event, with a major impact upon the real economy and a role in the pendular dynamics of capitalism. The crisis is endogenously generated by the way in which technological revolutions are assimilated into economic development. On the basis of recurring historical patterns, Perez introduces the major bubble collapse as marking the end of a job done by finance, and signalling the need for a structural shift with the return of an active State. During past decades, market competition and finance have installed a vast technological potential that can only yield maximum growth and social welfare if enabled by adequate institutional innovation. Recovery will require not only appropriate financial regulation and a redesign of its architecture, but also the conditions for promoting and profitably funding a structural change in the economy. The author discusses the elements for fostering a sustainable global golden age as being within the range of the possible. The move towards it, however, requires a combination of Keynes and Schumpeter.

The chapter "Innovation, Financial Activities and the future of the EU" by Pier Paolo Saviotti fits precisely with Perez' reasoning. The author analyses in detail the European version of the economic crisis and gives a well-founded diagnosis of its causative factors, namely insufficient intra-European coordination and insufficient innovation-orientation in several member states. Without doubt, the major achievements of the European economies are the highly developed welfare states with their complementary relationships to the productive sectors. However, in order to maintain this decisive complementary relationship, the competitiveness of the European economies requires a profound and comprehensive innovation-orientation. Even if this innovation-orientation – which demands, for example, a strengthening of education systems – requires some sort of austerity, the policy instruments are more easily enforced as abstinence today promises a better future. Most importantly, these policies are suited to calm financial markets which are – given an adequate time horizon – more worried about the absence of growth perspectives than about the speed with which public debts can be re-absorbed.

The chapter by Pelin Demirel and Mariana Mazzucato – "Innovation and economic performance (industrial and financial): Recent results and questions for future research" – surveys the empirical literature on the relationship between innovation and finance. The striking conclusion from a close survey of this literature is that there is almost no agreement on this relationship in either direction, namely the impact of innovation on finance or the impact of finance on innovation. In an aggregate perspective, market-based selection mechanisms do not adequately fulfil their role of rewarding innovative successful firms and punishing innovative failing firms. These rather specific relationships seem to depend upon a particular phase in the life cycle of a certain industry and are strongly influenced by all sorts of co-evolutionary relationships, which are responsible for delays, distortion of effects, and a crucial role for irregular distributions such as fat tails.

In their contribution "An Essay on the Emergence, Organization and Performance of Financial Markets: the case of the Alternative Investment Market" Valérie Revest and Alessandro Sapio give a sound overview of the evolution of European stock markets, with a particular focus on the financial requirements of technology-based small firms. They show the co-evolutionary relationship of the development of markets for finance and technological development and the influence of public regulation. In other words, the evolution of the financial markets impressively illustrates the complementary role of the three pillars in an economy.

The chapters by Pelin Demirel and Mariana Mazzucato, and Valérie Revest and Alessandro Sapio highlight the actor-based and sectoral dimension of the relationship between innovation with respect to firm growth and finance. Ryszard Wilczynski addresses macro-economic policy instruments and their ability to fight against economic crises. In his chapter – "Financing instruments to combat the crisis" – he makes the general case for stability of the international financial system before assessing different financial instruments according to their effectiveness in counteracting the negative effects threatening economic stability in situations of economic crisis. Financial markets turn out to be both facilitators of growth and vehicles of instability, leading to a deterioration in the balance of payments. Recent decades have been witnessing various sovereign liquidity to solvency problems of countries. To address these problems the author states that financial governance needs to be expanded, comprising innovative financing instruments. Among these instruments, an increasing emphasis on internationally coordinated, insurance-type tools is essential because only this will provide easier and more rapid access to liquidity.

The remaining three chapters change the angle of analysis, moving towards a sectoral and an international dimension. Gunnar Eliasson's chapter "Automotive dynamics in the Stockholm and Southern German regional economics – a comparison" treats a case which is of paramount importance for the Stuttgart area. The author embeds the sectoral development of the automotive industry into his theoretical framework of an *experimentally organized economy*. This notion is well chosen as it highlights the necessity of getting rid of the optimization principle and the short-term thinking which prevails in neoclassical economics when it comes to the analysis of innovation-driven economic development. The agents, endowed with heterogeneous competences, interact, and those interactions explicitly include not only the interactions between competitors in a market, but also those between manufacturing and finance actors. The decision logic of economic actors is best described as trial-and-error ("experimentally organized") and cannot follow any optimization calculus because of the uncertainty inherent to innovation.

The final two chapters of this volume deal with venture capital in Israel and Canada respectively. Of course, venture capital plays a decisive role in the financing of innovative companies – even if we have learnt from Pelin Demirel and Mariana Mazzucato that in venture capital industries the strong uncertainty stemming from innovation is wrongly processed by reducing time horizons, and that, in many cases, the real venture capital indeed comes from innovation policy programs. Israel is often characterized as the leading economy with respect to venture capital, and therefore a report on this important topic from an international expert from Israel was indispensable to the domain of innovation and finance. Morris Teubal, in his chapter "Innovation policy, innovation and innovation finance co-evolution: the Israeli case and some implications", also focuses on the co-evolutionary relationship which links innovation, innovation policy and innovation finance, and addresses the Israeli experience between 1969–2000. He concludes his highly informative study by suggesting possible implications of the analysis in relation to an innovation and structural change-led processes of economic development.

The final chapter of this volume, by Jorge Niosi, deals with "Venture capital in Canada". The Canadian case is of great interest because the recent success of the Canadian venture capitalists seems to be related to turning away from the US venture capital model and towards the Israeli model. In particular, Canada is moving from a massive backing to a more sophisticated support that bears resemblance to the Israeli Yozma Program. The venture capital industry – like other industries with high levels of risk and uncertainty – always requires public support. This support, however, is only a necessary and not a sufficient condition. As in other risk-intensive industries, a careful design of public support is indispensable.

In memory of Diogo Vasconcelos

One outstanding presentation of the Stuttgart Symposium was never submitted to be published in this volume. During the gestation process of this book one of the most creative members of the Lisbon Civic Forum, the strong supporter of Lisbon Civic Forum activities and strongly convinced European citizen Diogo Vasconcelos, tragically passed away, in July 2011. On behalf of the Lisbon Civic Forum we wish to express our sincerest appreciation and gratitude to Diogo, our creative friend, who we will keep in our memory always.

Notes

1 The Lisbon Civic Forum Symposium on "Innovation and Finance", which took place on 21 October 2010 in the Balcony Room of the Schoss Hohenheim, was generously supported by Stuttgart Financial, the Forschungszentrum für Innovation und Dienstleistung (FZID), and the University of Hohenheim.
2 Taken from the mission statement of the LCF (www.lisboncivicforum.org).

References

Braunerhjelm, P. and Feldman, M. (eds) (2007), *Cluster Genesis: Technology-Based Industrial Development*, Oxford University Press.

Dopfer, K. (ed.) (2005), *The Evolutionary Foundations of Economics*, Cambridge University Press.

Dosi, G., Freeman, C., Nelson, R., Silverberg, G. and Soete, L. (eds) (1988), *Technical Change and Economic Theory*, Pinter Publishers.

Erdmann, G. (1993), *Elemente einer evolutorischen Innovationstheorie*, Mohr Siebeck.

Fagerberg, J., Mowery, D. and Nelson, R. (eds) (2005), *Oxford Handbook of Innovation*, Oxford University Press.

Grebel, T., Pyka, A. and Hanusch, H. (2003), An Evolutionary Approach to the Theory of Entrepreneurship, *Industry and Innovation*, Vol. 10, Issue 4, 493–514.

Hanusch, H. and Pyka, A. (2007a), The Principles of Neo-Schumpeterian Economics, *Cambridge Journal of Economics*, vol. 31(2), 275–289.

Hanusch, H. and Pyka, A. (eds) (2007b), *The Elgar Companion on Neo-Schumpeterian Economics*, Edward Elgar Publisher.

Knight, F. (1921), *Risk, Uncertainty and Profit*, Hart, Schaffner & Marx, Boston: Houghton Mifflin Company, The Riverside Press, Cambridge.

Lucas, R. (1988), On the Mechanics of Economic Development, *Journal of Monetary Economics*, vol. 22(1), 3–42.

Nelson, R. and Winter, S. (1982), *An Evolutionary Theory of Economic Change*, Cambridge, Mass., Harvard University Press.

Pyka, A. (2002), Innovation Networks in Economics – From the incentive-based to the knowledge-based Approaches, *European Journal of Innovation Management*, Vol. 5, Issue 3, 152–163.

Pyka, A., Gilbert, N. and Ahrweiler, P. (2009), Agent-Based Modelling of Innovation Networks – The Fairytale of Spillover, in: Pyka, A. und Scharnhorst, A. (eds), *Innovation Networks – New Approaches in Modelling and Analyzing*, Springer: Complexity, 101–126.

Riuz, C. and Parra, C. (eds) (2013), *New Forms Of Organization In Knowledge-Based Societies: Social Innovation and Social Entrepreneurship*, Routledge.

Romer, P. (1987), Growth Based on Increasing Returns Due to Specialization, *American Economic Review*, vol. 77(2), 56–62.

Saviotti, P. P. (1996), *Technological Evolution, Variety and the Economy*, Edward Elgar Publisher.

Saviotti, P. P. and Pyka, A. (2004), Economic development by the creation of new sectors, *Journal of Evolutionary Economics*, 14(1), 1–35.

Schumpeter, J. (1911), *Theorie der wirtschaftlichen Entwicklung*, Duncker & Humblot, Berlin.

Solow, R. (1956), A Contribution to the Theory of Economic Growth, *The Quarterly Journal of Economics*, Vol. 70, No. 1, 65–94.

Sraffa, P. (1960), *Production of Commodities by Means of Commodities. Prelude to a Critique of Economic Theory*, Cambridge.

2 Financial bubbles, crises and the role of government in unleashing golden ages

Carlota Perez

Many analysts (in academic papers and books, journalistic articles, the news, and documentaries) have tried to identify the mechanisms that led to the 2007–2008 financial crisis, under the assumption that it was an unexpected "black swan" event.[1] The common idea is that there were extraordinarily risky and/or fraudulent behaviours that were induced or facilitated by regulation (seen as either insufficient or excessive, depending upon the author). The implication is that, by bailing out the banks, punishing the most notorious culprits and changing the regulatory conditions, the world will avoid another bubble and will return to business as usual. This article will argue that, however useful those analytical efforts may be for improving regulation, they remain superficial. It will hold that this was not an accidental collapse, but rather that it belongs to a special family of major once-in-a half-century boom-and-bust episodes, which are endogenous to the market system (Perez 2002), and that they regularly happen mid-way along the diffusion of each technological revolution. If this is so, then the recognition of this changes the nature of the solution to be sought. The world would not be facing a problem in the financial system, but rather in the real economy. It would not only be a question of creating conditions for healthy finance, but also of finding effective ways of reviving the economy, unleashing a new prosperity and making sure that the financial world plays a positive role in achieving that objective.

One of the reasons for the difficulty in seeing the underlying causes of the recent major bubble collapse is that standard economic theory is designed as a universal set of tools to function in the same manner at all times, assuming there is such a thing as "normal" circumstances. Yet, such an unchanging view of the workings of the market mechanism is based on a very narrow slice of social reality and on negating historical change. It is therefore unaware of the major upheavals that can be generated by specific technical changes, and while it chooses to trust the wisdom of markets, it tends to ignore the role of institutions, especially of governments.

To understand extraordinary times, however, one cannot do without interdisciplinarity. Only with a much wider framework, incorporating technology, institutions and their long-term historical interactions with the economy, is it possible to identify the fundamental regularities capable of explaining the causes and consequences of such major disruptions.[2] In the end, the adequacy of the solution to the financial crisis will depend upon the adequacy of the explanation.

The recurring pattern and its causes

The causes of these major financial booms and busts are to be found in the way that technologies evolve and are assimilated. Progress in market economies occurs by going through different successive Great Surges of Development driven by successive technological revolutions[3]. Human and institutional resistance to such radical changes results in capitalism experiencing pendular swings every two or three decades: from gilded ages to golden ages; from an initial installation period, through a collapse and recession, to a full deployment period.

After a gestation phase, sometimes driven by major government investments in the underlying new technologies (Mazzucato, 2013), an Installation Period begins. This is the turbulent process of creative destruction that replaces the old technologies with the new ones. It is a time of financial capitalism, characterized by unfettered free markets, intense competition and income polarization. The frenzied mood among investors and financial agents makes it possible to propagate the technological revolution and spread its new common-sense paradigm, but also to shift the excess funds that flood the market into a veritable casino, decoupled from the real economy. This culminates in a major financial bubble, the collapse of which marks the swing of the pendulum. What had, until then, been working for growth, profits and innovation is no longer effective. A recession in the real economy ensues.

The recessive interval lasts for a longer or shorter period depending upon the capacity of governments – consciously or intuitively – to establish an institutional framework capable of unleashing the installed potential. Deployment, then, is a period of production capitalism, aided by government, in order to fully deploy the new innovation and growth opportunities across the economy and to spread the benefits across society. When that potential begins to reach maturity and to face limits to productivity increases, to markets and innovation opportunities, the economy slows down. Conditions are then ready for the emergence of the next revolution and for the pendulum to swing back.

Before the current information revolution, the market system had experienced four similar technological upheavals, each diffusing in two periods with a panic and recession in between (see Figure 2.1).

The first great surge of growth was driven by the so called "Industrial Revolution" in England from the 1770s. After canal mania and the canal panic of 1793 there was the great British leap in the first decades of the nineteenth century. That was followed, from the 1830s, by the Age of Steam and Railways, which, after the mania, brought the railway panic of 1848 and, soon after, the Victorian Boom. The advent of cheap Bessemer steel, from the 1860s and 1870s, opened the way for a surge of innovation in the Age of Heavy Engineering –civil, chemical, electrical, naval – and for the first globalization. The panics that happened in Australia, Argentina and other Southern hemisphere newcomers hit the

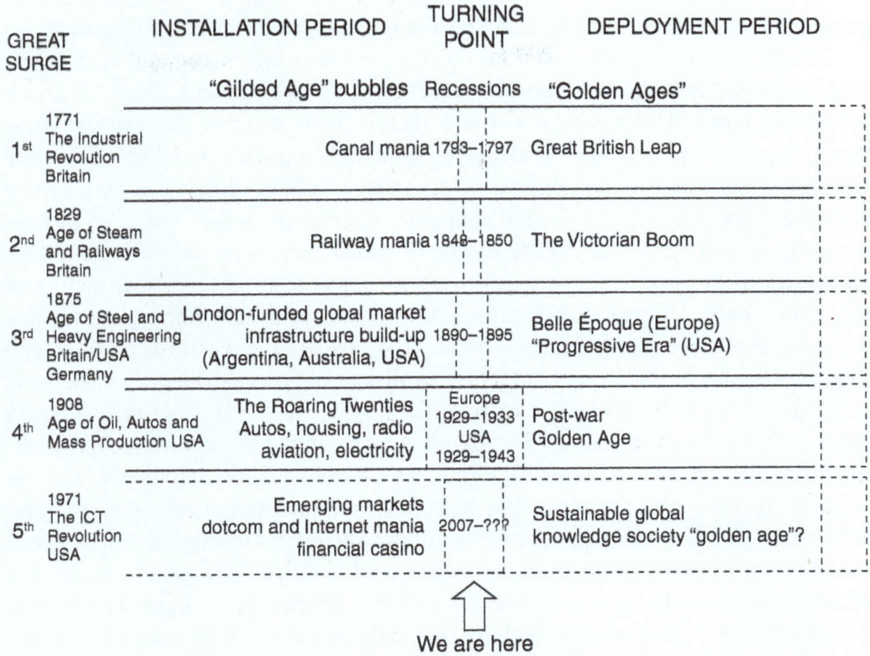

Figure 2.1 The historical record: bubble prosperities, recessions and golden ages (source: Perez and Carlota (2011) Fig. 1, p. 107).

promoters in the financial centres of London. The revival brought the Belle Époque in Europe and the Progressive Era in the US. In 1908 in the United States, a decade before the third revolution had reached maturity in Europe,[4] Ford's model-T inaugurated the Age of the Automobile and Mass Production. The great crash of 1929 ended the roaring twenties frenzy and led to the longest post-collapse recessive period to date: the 1930s. Resistance to the New Deal may be seen as one of the root causes of the prolonged stagnation. It took the experience of government–industry collaboration during World War II to enable acceptance of the full Welfare State and the Keynesian policies and institutions that facilitated the greatest economic boom in history.

In the early 1970s, when the potential of the mass production technologies approached exhaustion and markets became saturated, conditions were set for finance to search for other opportunities both in the global space and with the new microelectronics technologies. Once more, the installation of a technological revolution required the State to be moved aside in order to let the markets do the choosing, driven by high-risk finance. Now, after the double collapse of the NASDAQ in 2001 and the 2007–2008 bust, the pendulum is ready to swing back. Enabling policies are again necessary to unleash the deployment of the innovation potential created by the diffusion of the information and communications revolution (ICT). Power needs to be returned to production capital and a more patient financial world must be induced and encouraged in order to support it.

To understand why the assimilation process takes this shape and requires at first unfettered finance and then market-shaping by government, we need to ask why these constellations of radical new technologies warrant the term "revolution". Each of these Surges of Development encompasses and transforms the whole economy and is not limited to the new industries. Each can be called a revolution because it has a double character. On the one hand, it is a set of new products, new dynamic technologies and infrastructures with increasing productivity and decreasing costs that are therefore capable of explosive growth and structural change. Those are what most people will see as a technological revolution. On the other hand, each of them provides a new techno-economic and organizational paradigm that, together with the all-pervasiveness of the new technologies and the widening of markets by the new infrastructure, offers a quantum leap in productivity for all other activities and sectors. In practice, therefore, it will enable a massive process of rejuvenation. However, for the majority of existing companies the acceptance of such transformations is quite difficult. It is a complete change of "common sense" for competitiveness and a radical shift in best engineering and managerial practice. The natural resistance of all those that had been successful within the previous paradigm will require Schumpeterian "creative destruction", not only in terms of products and processes, but also in of behaviours and institutions. It is an intense process of learning the new and unlearning the old, for both producers and consumers.

The inertial forces resisting such profound transformations are at the root of the pendular swings. It is because the market system operates with two functionally distinct agents – financial and production capital – looking for profit in dissimilar ways that technological revolutions are assimilated in a sequence involving two different periods.

Production capital is the agent for accumulating wealth-making capacity. It is represented by the entrepreneurs and managers engaged in the production and distribution of goods and non-financial services. It is specific, fixed and knowledge-bound. By contrast, financial capital is infinitely flexible and mobile and is mainly moved by short-term criteria. It is the agent for reallocating and redistributing wealth, represented by investors, their banks and all financial intermediaries. It is in order to force the paradigm shift that financial capital takes over from production capital in controlling the direction of investment.[5]

The process that leads to the major technology bubble

When the surge of development and growth driven by a particular technological revolution approaches maturity, incumbent production capital becomes conservative: the most powerful companies and their managers are tied to their previous investment in fixed capital, to the specific knowledge of their markets, clients and suppliers, as well as to the technologies and strategies with which they had been successful until then. So even if innovation possibilities are drying up and market growth and productivity are slowing down, production capital is complacent and unwilling to face radical change. Financial capital then becomes restless and impatient for opportunities. In its search, it ends up in a tacit alliance with the new entrepreneurs to engage in a battle against the old paradigm. In doing so, the financial world will make a massive displacement of funds towards the new industries, while using all of its power to remove the obstacles posed to its freedom of movement by government and the established institutional framework. This behaviour establishes a huge market experiment to define the contours of the new – and renewed – industries and of the new paradigm through survival competition. It sees the astonishing growth of the new products and companies as well as the rejuvenation of the mature industries with the new paradigm (breaking the resistance through competitive pressures in unfettered free markets). The process eventually results in the emergence of new powerful companies and to the new sectors replacing the old giants as engines of growth in the economy.

The excitement of the extraordinary gains from innovation in the real economy leads to excess funds flowing into the stock market in the hope of participating in the easy-profits game. This produces the rapid inflation of the desired new stocks and initiates a major episode of Minskian instability (Minsky 1982). The extraordinary bounty induces the financial world to engage in all sorts of innovation to mobilise it: some good, some doubtful, some even fraudulent and illegal. Thus finance soon completely decouples from the new economy and adopts a casino-like behaviour. As Galbraith (1990–1994) remarks, this is all done with great confidence in what is seen by the financiers as their new power to "create wealth" by the strength of their genius. The result is a major bubble and its inevitable collapse.

Nevertheless, the consequences of these particular bubbles that occur at mid-surge driven by a radically new set of technologies are not all negative. By the time the collapse happens, the new industries and infrastructures will have been fully installed in the territory and the new paradigm will have become the new common sense for innovation and competitiveness. By then, the new and renewed production capital will be better able to lead the economy and to decide on investment. Installation has been achieved; deployment can now be enabled; finance capital must cede the leading role to production capital.

In essence, the sequence installation-bubble crash-deployment describes the process of assimilation of a technological revolution and the full reaping of its fruits. Installation is a period of creative destruction, wielding the power of the new, that can be seen as "supply-push", while Deployment is a time of expansion, growth and innovation across all industries – aided by the new technologies and their paradigm – and driven by "demand-pull". In the first, innovation is concentrated in the new industries and finance; in the second, it is the whole world of production and the institutional sphere where the innovative forces are at play. The first is typified by income polarization; the second tends to reverse that process and leads to a better distribution of well-being. Finance must make the difficult shift from self-serving casino to funding the expansion of the real economy and sharing in its profits, and from putting pressure on production companies to yield short-term gains to serving their needs and longer-term projects; in other words, from impatient to patient capital.

Obviously, that change of attitude and transfer of power is never easy. It requires government intervention to radically modify market conditions, incentives and regulation in order to reorient financial capital towards supporting the real economy by making it more profitable to do so than to continue in the casino mode. This entails extraordinary political leadership to confront the *entente* between financiers and politicians that is usually woven during the fantastically profitable times of the bubble. It also involves confronting the market fundamentalism that accompanies Installation and makes unfettered free markets appear as the reason not only for the bubble prosperity, but also for prosperity in capitalism at any time.

In fact, both pendular shifts are times of political and ideological confrontations. At the end of Deployment, stagnation has to be deep enough and long enough to bring back the dynamics of the market and eliminate the institutional obstacles to the diffusion of the new paradigm. At the end of Installation, the financial collapse and its consequences have to be big enough to weaken the power of finance and bring back the regulatory power of the State, the long-term interests of production capital, and the welfare interests of the public.

The double bubble at the turn of the century

A unique feature of our time is that the major mid-surge panic happened in two episodes: first, there was the collapse of the NASDAQ at the end of the internet mania in the 1990s. That bubble was driven by technological innovation centred on ICT. Second, there was the 2007–2008 meltdown. In this case it was the massive wave of financial innovation using ICT that drove the easy credit bubbles with high-risk shadow banking and the sub-prime madness in the housing market. Understanding the continuity of the double bubble is crucial for identifying the nature of and the solution to the current crisis (Perez 2009).

The bubbles of the 1990s and the 2000s are simultaneously linked and different. They both concentrated innovation and asset inflation in technology and finance, but they did so in opposite proportions. Together, technological and financial innovation represented more than half of initial public offering (IPO) activity in US stock markets (see Figure 2.2), reaching as much as 70% at the peak of the major technology bubble. Yet it is clear that, in the 1990s, the technological offerings prevailed by far, while it was finance that was the more active in the 2000s.

The IPO activity of these two sectors was followed by rapid market asset inflation. Indeed, in terms of market capitalization, these two sectors had a very pronounced bubble-like behaviour. As a consequence, at the peaks of the two bubbles they jointly represented over $5 trillion dollars in market value (see Figure 2.3), which was about 35% of the total stock market. It should be noted, though, that the two sectors had different rhythms of inflation: the ICT stocks

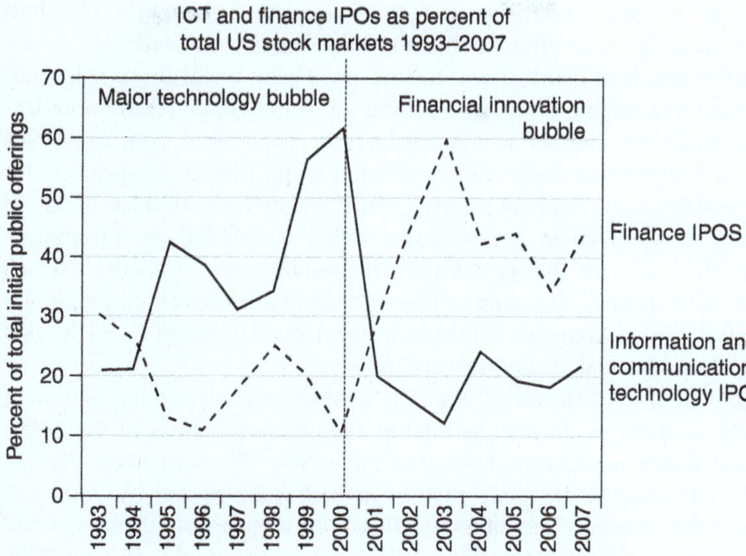

Figure 2.2 The 1990s and 2000s: a switch in the composition of the new offerings (source: Thomson).

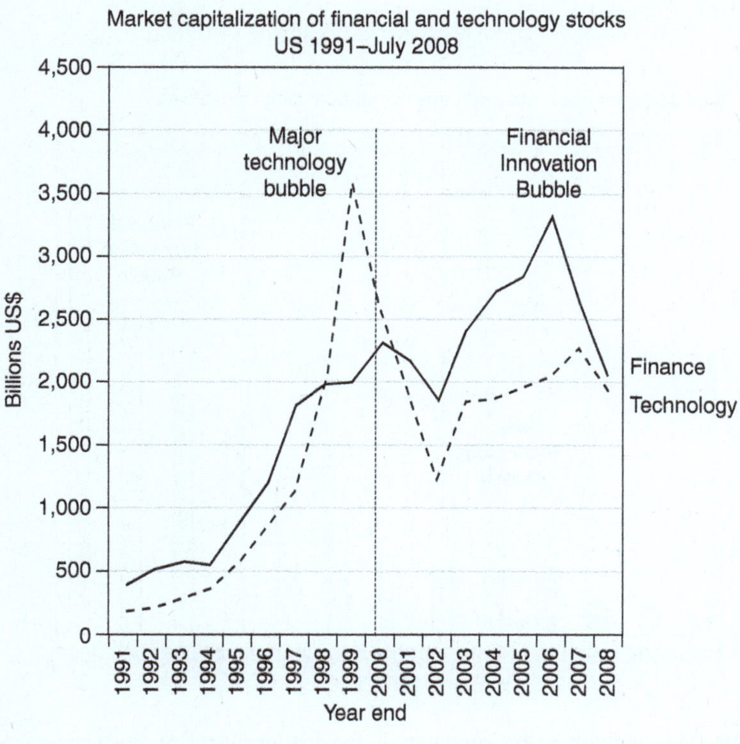

Figure 2.3 The 1990s and the 2000s: a switch in differential asset inflation (source: Thomson).

were valued at more than $3.5 trillion with the financial ones being just over half of that. In the financial innovation bubble, the proportions were reversed.

It must be emphasized that ICT and finance have been intensely interdependent throughout these bubble times. The setting up of the global telecommunications infrastructure for internet would not have been possible from the 1990s without capital gains in the stock market to foster major fibre optics projects for full coverage without dividends or profits. In turn, the housing bubbles could not have been so intense without securitization (which depended on information technology) and without the possibility of trading them globally (which depended on the internet). The same can be said about derivatives, credit default swaps and all the other high-risk synthetic instruments developed in this period with the help of sophisticated computer software.[6]

So, in spite of their differences, the two bubbles are intertwined and were fundamentally continuous. In fact, several of the casino practices of the 2000s were an intensified continuation of those of the 1990s. The derivatives market, for instance, increased at the same rhythm in both bubbles and by 2007 had reached a notional amount of about $380 trillion (see Figure 2.4). This astonishing figure represents seven times the GDP of the whole world. But by 2000 derivatives were already built upon values equivalent to global GDP.

Another aspect of the continuity between the two bubbles is the intense bias towards financial profits in the corporate sector. Whereas in the early 1990s the

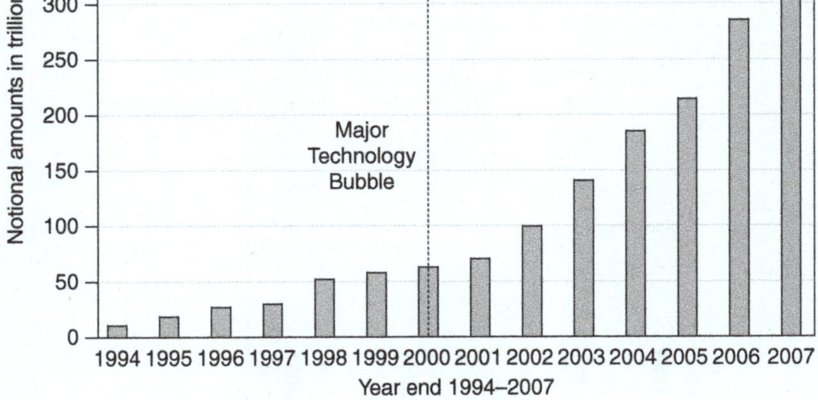

Figure 2.4 The 1990s and the 2000s: continuity in the instruments of casino-type speculation (source: International Swaps and Derivatives Association, Inc).

financial and non-financial corporate sectors still saw their profits increase apace with one another and with GDP, from the mid-1990s onwards profits in both sectors outpaced the growth of the economy, but with a much more pronounced increase in those of financial companies (see Figure 2.5). This phenomenon, which was already significant at the peak of the NASDAQ bubble, reached extreme proportions in the 2000s.

It is interesting to note that the profits of non-financial corporations – which could be taken to represent the real economy – grew at pretty much the same rhythm as GDP throughout the 1990s and until 2003. From then on they seem to decouple from the economy and grow as fast as the financial sector profits (which had decoupled from the rate of growth of GDP in the early bubble). The result is that they both more than doubled their benefits in the four years between 2003 and 2007. However, this may not be a reflection of an outstanding performance by the production companies in terms of sales and productivity. Unfortunately, what it probably indicates is the "financialization" of the real economy. According to Krippner (2005), it was the financial arm of the non-financial corporations that was the source of such extraordinary profits. Thus, what we have witnessed is a complete decoupling of profit-making from real production of goods and services. It is the setting up of a massive casino operation encompassing not only the stock market and the banks – shadow or otherwise – but also the agents of the "real" economy (Lazonick 2010).

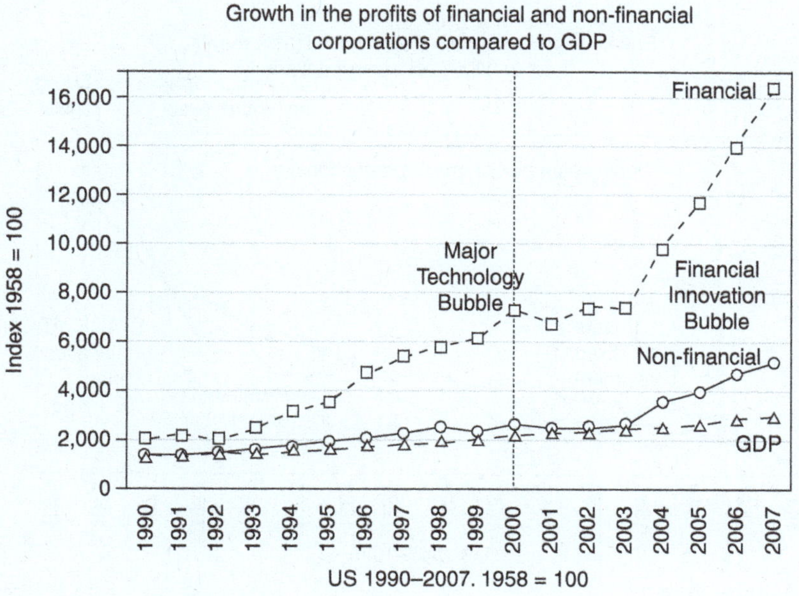

Figure 2.5 The intensified bias towards financial profits (source: Bureau of Economic Analysis).

From installation to deployment

The world is now at the turning point. The crash of 2007–2008 has plunged the real economy into a recessive mode and has revealed the decoupling of finance from real wealth creation as well as the polarization of income and the illusory growth that was hidden behind the frenzied bubble years. The time has come to move from the gilded prosperity at the end of the installation period to the truly golden prosperity of the deployment period. This will require a very substantial shift in the market context so as to orient the behaviour of the investment agents.

A world in deployment is very different from a world in installation. The experience of the previous deployment, in contrast with the recent installation, can serve to illustrate the difference between the two periods. In particular, one can observe the change in the relative behaviour of direct investment and financial flows as a proxy for the leadership in investment.

In the deployment period of the previous surge, from 1947–1974 fixed investment consistently outpaced financial credit flows (see Figure 2.6), whereas during the Installation of the current fifth surge (1970s–2000s), finance increasingly decoupled from investment in the real economy (see Figure 2.7).

Yet historically these major bubbles have played a crucial role in the market economy: they enable the installation of the bases for the future. In their wake, there is generally in place enough of the new infrastructure for a decade or more; new production and consumption models have been established, the new paradigm is accepted by all as common sense, and the economy counts upon new entrepreneurial giants as leaders of the new industries, which can act as the new

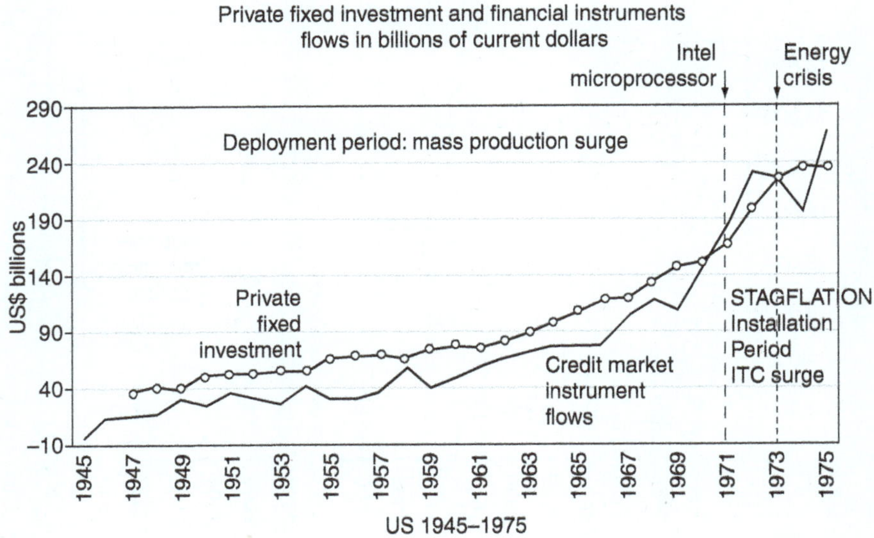

Figure 2.6 In the deployment period of the previous surge: 1947–1974, Fixed investment outpaced financial credit flows (source: data from BEA and Fed, period indications by the author).

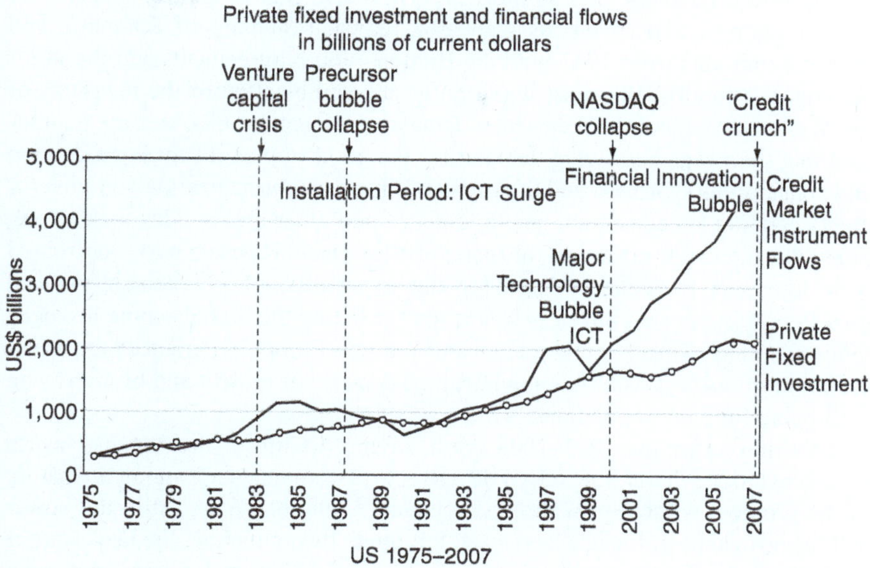

Figure 2.7 During the installation of the current fifth surge 1970s to 2008 finance increasingly decoupled from investment in the real economy (source: data from BEA and Fed, period indications by the author).

engines of growth for the next period. Essentially, there is vast new potential for using the ICT paradigm to innovate across all industries and activities and this would have been difficult to achieve without the capital gains and the financial excitement of bubble times.

In other words, these major boom-and-bust periods prepare the economy for full expansion with the new technologies and their new common-sense paradigm. However, as discussed above, the unleashing of a healthy deployment period requires shifting the control of investment from financial to production capital, from short-term to longer-term decisions, from quick capital gains to patient capital. Achieving this power transfer demands an active come-back of the State in order to radically reshape market conditions and profit opportunities away from casino activities and in favour of the real economy. That implies a clear understanding of the problem at hand and bold institutional creativity.

At the turning point, after each of these major crashes, governments face three main tasks. The first is to rapidly perform "intensive therapy" for the financial world. The second is to thoroughly examine and redesign financial regulations and architectures. The last – and very far from least – is the induction of a structural shift in the real economy that will reshape market conditions to fully exploit the installed innovation and growth potential for the benefit of all.

The last time around – the 1930s – many mistakes were made that led to the depression. But, in the end, the revival was engineered by a bold set of policies: bank regulation (including the Glass–Steagall Act in the US, the separation of savings and investment banks almost everywhere, exchange control measures in many countries, and so on) as well as a radical reshaping of economic and market conditions from 1943 with the Bretton Woods agreements and the international institutions. But most importantly, the combination of the measures of the Welfare State and the Keynesian demand management mechanisms guaranteed that there was a growing demand for the goods typical of mass production and consumption. Growth and innovation followed suburbanization and government spending (in education, health, defence and other areas). This took a long time after the major crash and, at first, the New Deal proposals were confronted with ferocious resistance. Yet, after the experience of government-induced growth during the war, business was ready for letting the State become involved in the economy. Thus North America and Western Europe experienced the greatest boom in history, with two uninterrupted decades of growth and of increasing well-being for their populations.

This time, after the 2007–2008 crash, saving the life of the financial world was so overdone that it transferred the crisis to governments, changing the nature of the problem without providing a sustainable solution. Worse still, the power and the arrogance of finance have been left intact by not making the banks suffer enough of the losses warranted by their irresponsible risk-taking and even wrongdoing. This power makes them ready and able to block the necessary measures involved in the other two tasks.

In the current globalized world, the task of redesigning regulation and financial architecture cannot be a national matter. This time global finance needs both better national regulation and a global regulatory "floor" that will avoid competition to the bottom. The great difficulties experienced by the regulatory attempts in every country are witness to the expectation of the over-rescued financial world to maintain its unreasonable claims to high profits in the midst of stagnant economies as well as to block any attempt at supranational regulation.

As to the necessary structural shift in the real economy, the basic problem is that it is not even on the table. There is a general belief that since the crisis was created by finance, it is enough to save the financial world in order to go back to "business as usual". In addition, the way in which government deficits are being confronted is with a view to rescuing the banks at any cost. However, "healthy" banks in a sick and languid economy inevitably become – or remain – casino banks. The real solution to the problems of growth, employment and finance at the present turning point is to revive the production economy by radically reshaping market conditions in order to encourage innovation, investment and employment creation with healthy profit prospects. If this were achieved, the financial world would reap its profits not from manipulations and gambling, but as a share of the real wealth created by the production world with the support of finance. The task of reshaping the market through a modern sort of industrial policy is rarely considered to be the role of governments at this time, but recovery will be very difficult without it.

A global sustainable Golden Age ahead?

The post-war Golden Age was unleashed by leaving behind the free market policies of the roaring 1920s and the recessions of the 1930s. Finance was regulated favouring real investment and income was redistributed to improve demand profiles for suburban living and mass consumption. It was the achievement of Western democracies to set up a positive-sum game between business and the great majorities (as workers and consumers).

Is a new positive-sum game possible today? Our answer is yes. But it will have to be both global and national; it will need to be sustainable in environmental and social terms and it will involve making the best use of the all-pervasive innovation potential provided by the information and communications technologies (ICT) and their techno-economic paradigm. In a few words, the formula would be combining ICT with "green" growth and full global development.

Full internet access at low cost for all is equivalent to electrification and suburbanization in terms of facilitating innovation and investment as well as widening demand. This time, access to ICT also serves to educate the future labour force and to shape the patterns of consumption towards intangibles and creativity.

"Green" growth (not zero growth) entails revamping the transport, energy and production systems to make them sustainable, and is equivalent to post-war reconstruction and suburbanization in terms of job creation. For those who have lost manufacturing jobs, it can provide employment not only in the new green products but also in maintenance and recycling, while it can also multiply the productivity of scarce natural resources.

Full global development involves incorporating successive new millions into sustainable consumption patterns and is equivalent to the Welfare State and government procurement in terms of demand creation. If products are to be made durable, upgradeable, reusable, maintainable and recyclable, producers will need constant waves of new consumers. Those waves would be provided by the emerging and the lagging countries as they develop. Moreover, it is not feasible to incorporate such masses of new consumers without "green" patterns of consumption (we only have one planet).

The technological revolution and the global boom have provided the wealth-creating potential for a sustainable global Golden Age. The challenge is to collectively build positive-sum games between business and society, between the advanced and the advancing countries and between humanity and the planet. But the goals of policy need to be clear for all of the agents involved.

The goal of new regulation is not to constrain finance, but to reorient it. It should become more profitable to fund expansion and innovation in the real economy than to "play" in the casino of synthetic instruments and derivatives. The goal of stimulus is not just to put money into the economy, but to shape the demand opportunity space, making it profitable to innovate in agreed convergent directions. We need a modern industrial policy in a global context

Market fundamentalism (and the sort of narrow economics theory that supports it) is one of the main obstacles to unleashing a healthy deployment. But invoking Keynes is not enough; we also need Schumpeter, together with a disposition to make bold institutional innovation on local, national and global spaces.

Notes

1 Two examples in this wide spectrum are Taylor's (2009) academic analysis, ascribing the boom and bust to loose monetary policy (excessively low interest rates) and Charles Ferguson's *Inside Job,* a full-length documentary concentrating on the bad practices of the financial world. From very different points of view, they both argue that the crisis was avoidable through policy action.

2 This much more encompassing sort of economic theory is what Evolutionary Economics proposes in order to enrich our understanding not only of such extraordinary times, but also of how the real economy works. See, for example, Dosi *et al.* (1988), Freeman and Louçã (2001), and Hanusch and Pyka (2007).

3 For technological revolutions as drivers of growth in capitalism, see Schumpeter (1939). For a proposed explanation of such discontinuities in technical change, see Perez (2002, pp. 27–32).

4 Such overlaps, as well as the dating of the revolutions, differ from that of Schumpeter (1939) because the notion of "great surges of development" (Perez, 2002) focuses on the irruption and gradual assimilation of each technological revolution rather than on the long upswings and downswings in GDP that Schumpeter defined as "long waves".

5 In the current crisis, an additional problem is that production capital has been losing its traditional identity by adopting the short-termism and some of the other practices of financial capital, so that in this case part of the solution would be to re-establish a clear division of functions between the two (Mazzucato and Shipman, 2012). The overemphasis on short-term stock prices was such that production companies often sacrificed their human capital as well as their R&D spending in order to do stock repurchases to boost share prices (Lazonick and Tulum, 2011).

6 As Kregel and Burlamaqui (2005) noted, "financial innovations that facilitate the financing of innovation in business tend to decrease transparency concerning the risks being borne in the system, raising the possibility of ever-increasing financial risks and ever-decreasing understanding of the extent of these risks."

Bibliography

Dosi, G., Freeman, C., Nelson, R., Silverberg, L. and Soete, L. (eds) (1988). *Technical Change and Economic Theory*, London: Francis Pinter.

Freeman, C. and Louçã, F. (2001). *As Time Goes By: From the Industrial Revolution to the Information Revolution*, Oxford and New York: Oxford University Press.

Galbraith, J. K. (1990–1994). *A Short History of Financial Euphoria*, New York: Whittle-Penguin.

Hanusch, H. and Pyka, A. (eds) (2007). *The Elgar Companion to Neo-Schumpeterian Economics*, Cheltenham: Edward Elgar.

Kregel, J. and Burlamaqui, L. (2005). "Banking and the Financing of Development: A Schumpeterian and Miskyan Perspective", in Dimsky, G. and de Paula, S. (eds), *Reimagining Growth: Toward a Renewal of the Idea of Development*, London: Zed Books, Ch. 6.

Krippner, G. R. (2005). "The Financialization of the American Economy", *Socio-Economic Review*, vol. 3, 173–208.

Lazonick, W. (2010). "Innovative Business Models and Varieties of Capitalism: Financialization of the US Corporation", *Business History Review*, vol. 84 (Winter): 675–702.

Lazonick, W. and Tulum, O. (2011). "US Biopharmaceutical Finance and the Sustainability of the Biotech Business Model", *Research Policy*, vol. 40, no. 9, 1170–1187.

Mazzucato, M. (2013). *The Entrepreneurial State: Debunking Private vs. Public Sector Myths*, London: Anthem.

Mazzucato, M. and Shipman, A. (2012). "Finance for Creative Destruction vs. Destructive Creation", FINNOV DP 2.10.

Minsky, H. (1982). "The Financial-instability Hypothesis: Capitalist Processes and the Behavior of the Economy", in Kindleberger, C. P. and Laffargue, J. L. (eds), *Financial Crises. Theory, History and Policy*, London, New York: Cambridge University Press, pp. 13–39.

Perez, C. (2002). *Technological Revolutions and Financial Capital: The Dynamics of Bubbles and Golden Ages*, Cheltenham: Elgar

Perez, C. (2009). "The Double Bubble at the Turn of the Century: Technological Roots and Structural Implications", in *Cambridge Journal of Economics*, vol. 33, no. 4, 779–805

Perez, C. (2011). "The Advance of Technology and Major Bubble Collapses", in Linklater, A. (ed), *On Capitalism,* Stockholm: Ax:son Johnson Foundation, pp. 103–114.

Schumpeter, J. A. (1939[1982]). *Business Cycles* (2 vols.), Philadelphia: Porcupine Press.

Taylor, J. B. (2009). "The Financial Crisis and the Policy Responses: An Empirical Analysis of What Went Wrong", *NBER Working Paper* 14631, Cambridge, MA, National Bureau of Economic Research.

3 Innovation, financial activities and the future of the EU

Pier Paolo Saviotti

Introduction

This chapter was written after a meeting which took place in October 2010 in Stuttgart about the above theme. By that time the crisis of the world economy that is still with us was already underway, but the crisis of the Euro area was not as pronounced as it has subsequently turned out to be. That an economic crisis of the dimensions of the present one induced a crisis of the Euro area cannot *ex post* be considered surprising. The Euro area did not satisfy the requirements for a common currency area, and the fact that a similar crisis did not occur before is proof that it was a fair-weather currency (Rose, 2006; EIU, 2011). One has only to hope that the creators of the Euro understood the missing parts of the scheme and planned to create the required conditions later. If interpreted in this way the construction of the Euro has to be considered as a project underway which needs several further steps to be completed. Interestingly, the present crisis has already induced a series of responses which, while not going far enough, are moving in the right direction. In fact, most of the required conditions amount to a degree of renunciation to national sovereignty of the member states and to the adoption of more advanced coordination rules. The present crisis has pushed the Euro area towards a dichotomy: either move towards a higher level of supranational integration or accept the collapse of the Euro with all the consequences that would entail. The measures adopted so far go in the direction of a higher degree of integration but stop short of what would be required to transform the Euro area into an adequate common currency area. This chapter will contribute to the discussion of the future development of the Euro and of the EU by focusing on the following ideas:

- The measures adopted so far are exclusively financial and do not take into account other aspects of the economic systems of the member countries.
- The measures adopted so far show an excessive emphasis on austerity at the expense of growth.
- The measures adopted so far can only be temporarily successful but, even if successful, they are likely to lead to a growing divergence of the national economies of the member countries, thus undermining the stability of both the Euro and the EU.

- To ensure the future development of the EU and of the Euro a higher degree of coordination of economic and innovation policies is required.

While these considerations may seem out of place in a situation of great urgency, to focus exclusively on measures designed to react to the emergency without any attention to longer-term problems may turn out to be extremely futile, on account of asking EU citizens to make great efforts only to find out a little later that they were misplaced and insufficient. Crises can be sources of suffering, but also of renewal. It would be very unwise to react to this crisis by focusing exclusively on the measures needed for immediate survival without trying to find a solution designed to create an adequate place for the EU in the emerging economic and political structure of the world. It is much better to develop a creative response to this crisis by planning the future steps required to make the Euro area more competitive and capable of adapting to the new international environment in which it has to live. Financial markets are not only worried about the reduction of the present debt, but also about the perspectives of the Euro area. To create a credible development plan which combines growth prospects and debt reduction could convince both financial markets and European citizens that, if sacrifices are required, they will be an investment for a better future and not an irrecoverable loss.

The EU

In this section some of the most important aspects of the European project will be recalled. The creation of the EU (then European Economic Community) in the 1950s was, to a considerable extent, determined by the consequences of the two world wars. If the period between the sixteenth and nineteenth centuries had witnessed the creation of several European nation states, the twentieth century showed what destructive outcomes the confrontation of such nation states could lead to. Such a situation showed that there was a need for inter-country coordination mechanisms. The destructive effect of those wars could be interpreted as the failure of the European nation states to find a form of inter-country coordination that could ensure peace and prosperity. The initial European treaty (*Official Journal C, 2006; Treaties of Rome*) was due to a non-negligible extent to the objective to bind Germany and France together in order to ensure that they would not wage war against each other in future.

From the very beginnings of the EU project a tension existed between two concepts of the EU, which can be described by the dichotomy Federation vs Confederation (Josselin and Marciano, 2006). Whereas in a federation nation states would abandon a large part of their sovereignty, in a confederation they would preserve a large part of it. This same tension could be perceived as that between a purely economic concept of the EU, within which customs barriers would be eliminated and one large common market would be created, and a more political conception, within which nation states would release a large part of their sovereignty and construct some novel forms of inter-country coordination, possibly ending with the constitution of a federal Europe. In this sense the EU project is one of the greatest institutional innovations of the second half of the twentieth century, and the development of the EU has to be considered an experiment in the construction of tools and mechanisms of supranational coordination.

Another extremely important component of the EU project is the balance between competitiveness and cohesion (Sharp, 1998; Clarysse and Muldur, 2001; McCann and Ortega-Artiles, 2011, see Competitiveness Cohesion refs). At the centre of this is the vision of an EU capable of being both competitive and able to provide its citizens with a high degree of social justice and material welfare. This combination has sometimes been referred to as the EU social model. While some of the values underlying the construction of welfare states in EU countries can be largely shared in member states, recent research shows that in reality there are at least four EU social models (Esping Andersen, 1999; Sapir, 2005; Lefevre and Meda, 2006). The nature of such social models, as well as that of the national economies, constitutes clear evidence of the persistent heterogeneity of the EU. While such diversity could be a source of strength for the EU it is also a cause of persistent difficulties. As will be argued in the following paragraph, the relationship between competitiveness and cohesion is not a simple adversarial one, but is characterized by positive or negative feedback loops between the two. In this sense the diverging competitiveness which we have recently observed amongst EU member can be extremely destabilizing and can undermine the existence of welfare systems.

Any judgment about the EU project would inevitably depend upon (i) the assumptions involved and (ii) the changes that the whole EU project and its external environment have undergone since the 1950s. A convinced federalist would probably regret the limited degree of supranational coordination achieved so far, while a supporter of a purely economic conception of the EU would complain about the excessive inroads made by the European Commission in national legislations. Whatever these possible differences in judgment, the ability of the EU to attract new member states is a clear demonstration that potential applicants have a positive perception of the EU project. However, it is important to bear in mind that the external environment within which the EU project is carried out has undergone considerable changes since its beginning.

In reference to the initial objectives of the EU project, one might conclude that the need to bind together EU countries in order to avoid any future war has been achieved. A fundamental question would then arise about the nature and extent of new objectives which could have been induced by the changes in the EU external environment. The EU was created during a period in which the world was dominated by two superpowers relying upon irreconcilable ideologies and locked into a cold war. Some commentators have even stressed the need for protection from the Soviet Union as a further inducement to the creation of the EU. With the end of the cold war it briefly seemed as if the EU could become part of a triad (i.e. the USA, the EU and Japan) which would ensure stable governance of the world. The subsequent and absolutely remarkable emergence of

China and possibly of India, following the trail blazed by South Korea and Taiwan, is causing what is probably the greatest change in world economic and political power since the 1930s. Like the 1929 one, the present world crisis is not purely financial, but it signals a very profound change in the power structure of the world. Such a change in the external environment has profound implications for the EU project. In particular, the balance between competitiveness and cohesion is heavily challenged. Some commentators stress the high cost of the welfare state and its negative impact upon EU competiveness. In this view the welfare state is seen as competing for resources with the competitiveness of national economic systems. While this approach contains some grains of truth, it exclusively stresses the adversarial nature of social assistance and of economic activities. In fact, advanced welfare states only emerge when countries overcome particular thresholds of economic development. This was the case for the European countries and a recent confirmation of this trend is provided by the emergence of welfare states in the emerging Asian countries (*The Economist*, September 8, 2012). In fact, the various EU welfare states are the source of a non-negligible share of economic activities. Even private consumption can be enhanced by the presence of a well-designed system of social assistance, including, for example, subsidizing schooling, health care and pensions. The present rebalancing efforts of China from exports to internal consumption (*The Economist*, May 26, 2012) provide clear proof that a welfare state can be both a substitute and a complement for purely economic consumption. If welfare states only emerge in countries which have achieved a minimum level of economic development, then they cannot be considered an expensive form of consumption which can be dispensed with during periods of crisis. The formation and size of welfare states contributes to the output and the growth of countries. For example, the health and education systems of advanced countries provide a very large share of national employment and output. From another point of view, the absence of a well-developed welfare state would considerably reduce the internal demand of a country due to the high savings that citizens would need to make in order to educate their children or to be able to afford medical care. However, it is not possible to imagine that a national economy can be transformed into a gigantic welfare state. Without turning to the very difficult question of the precise role of the welfare state within the mechanism(s) of economic development of countries, we can state the following:

- Welfare states emerge only in countries which are above a given threshold of economic development.
- Welfare states can contribute to the output and to the growth of countries but cannot *substitute* other sectors of economic activity. Let us call these other sectors the productive sectors of the economy, for want of a better term. One property of these productive sectors is that they have to be able to contribute to exports.
- When an advanced economic system works well the welfare state can be *complementary* with respect to the productive sectors. In this case there is

co-evolution of the welfare state and of the productive sectors. A co-evolutionary relationship can both enhance the process of economic development when the feedback between the welfare state and the productive sectors is positive, but can also harm development when such feedback is negative.

- An example of negative feedback can be given by the combination of an increase in the costs of the welfare state (for example, due to population ageing) and of the falling competitiveness of the productive sectors, leading to a negative balance of payments and inducing a need to borrow more (possibly on international financial markets), leading in turn to greater indebtedness and ending with a vicious circle of growing costs and indebtedness combined with falling resources.

As part of the ongoing process of globalization we are now simultaneously observing European welfare systems under pressure, due partly to falling competitiveness, and emerging countries reaching the conditions under which welfare systems are advantageously introduced. Of course, welfare states are at the centre of all EU social models, although in different ways. In particular, they are central to the maintenance of an adequate balance between competiveness and cohesion. Thus, the nature of the EU social models and its interaction with competitiveness deserves greater attention than it has been given so far.

Whatever conclusion one may come to about the above problem, there is no doubting that competitiveness is an important component of the stability and development of the EU. Emerging countries such as South Korea, Taiwan and China are making inroads into a number of industries and technologies which were the preserve of the USA or of European countries. Even admitting that the welfare states are not substantially inimical to the competitiveness of the EU, it would seem very unwise to conclude that one does not need to worry about competitiveness. The complementarity also works in the direction that a welfare state can only be supported within a well-functioning economic system. What is at stake is not the existence of welfare states and social models, but the way in which they are organized. In particular, depending upon their organization, social models can favour or inhibit innovation, structural change and growth. In the kind of world which is taking shape now, the innovation capabilities of EU countries are crucial in order to maintain a welfare state and to ensure a high degree of social cohesion.

We will return to the role of innovation in the EU later in this chapter. However, now we need to discuss one of the most important events which, together with the present world crisis, are exercising an extremely powerful influence upon the stability and the future development of the EU. This event, which will be discussed in the next section, is the creation of the Euro and of the Euro area.

The euro

The adoption of the euro as a common currency was one of the most important steps in the development of the EU. There can be no doubt that the countries of the EU did not satisfy the conditions required for a common currency union (see Mundell, 1961,1968; Rose, 2006; EIU, 2011). For example, the Euro area did not have a single bank regulator, a lender of last resort, or a credible set of fiscal rules (Eichengreen, 2011). However, at the beginning the euro was surprisingly successful, probably due to the favourable conditions that allowed some observers to consider it a fair-weather currency. The crisis which started in 2008 began to reveal the weaknesses of the Euro area. A division rapidly emerged between northern and southern European countries, based on their different approaches to fiscal policy (De Grouwe, 2010, 2011). By reproducing a now widely diffused stereotype, we could say that the division in the Euro area was between "wise" northern countries who managed their finances well, and "dissolute/lax" southern countries who had accumulated unsustainable debts. Such a situation clearly showed the weakness of the euro, consisting of the lack of a central banker. The ECB (European Central Bank) sets one common interest rate for all the Euro-area countries, thereby forcing "weak" countries to accept an interest rate which is not suited to their conditions. Outside the Euro area the member countries would have been able to restore their competitiveness by devaluing their currency; within the Euro area this option is forbidden and countries have a much more limited range of available measures for restoring competitiveness, including the politically unpalatable one of reducing wages. Such defects did not appear before, but have become cruelly evident since the beginning of the present crisis, threatening the existence of the Euro area and the EU.

The problem of the euro is not one of indebtedness, or at least one of *average* indebtedness, but of governance (De Grauwe, 2010). As long as budgetary policies (spending and taxation) remain vested in the hands of national governments and parliaments, the political responsibility for decisions about spending and taxation also rests with these national governments and parliaments. The latter face political sanctions by national electorates. Neither the European Commission nor the other members of the Council face the political responsibility for the measures they impose upon a member country: "No taxation without representation belongs to the essence of democracy" (De Grauwe, 2010).

Thus, the crisis showed that the euro project lacked a political dimension and that to acquire such a political dimension meant moving towards a higher level of integration of the national economies of the member states. The measures adopted so far – such as the EFSF, the ESM, and, in particular, the recent decision by the ECB to buy unlimited quantities of government bonds of member countries, christened as Outright Monetary Transactions (OMT) – move in that direction and represent a considerable change in the functions of the ECB, making it more like a lender of last resort. However, the process of integration necessary to create the level of supranational integration needed to transform the Euro area into a coherent economic entity is very complex and unlikely to be completed soon, even under the strong inducements provided by the present crisis. Examples of the difficulties involved in pursuing the process of European integration needed for the survival of the Euro area are the obstacles encountered by the creation of Eurobonds of a European Banking Union. At this stage it is still possible for the Euro area to collapse, although the probability of its survival has been considerably enhanced by the measures so far adopted. However, these measures are at best necessary, but not sufficient conditions to stabilize the Euro area. Even if they were immediately successful they would at best avoid a growing indebtedness of "weak" countries, but at the expense of making them relatively poorer with respect to "strong" EU countries. Whilst seeming immediately successful, such a solution is much more likely to destabilize the Euro area and the EU in the medium to long term. In fact, the solution of the problem depends not only on reducing the indebtedness of the weak members of the Euro area, but also on improving their competitiveness and growth prospects. Of course, the timing of each measure is different. The measures applied or planned so far are urgently required in order to prevent the collapse of the Euro area, but they cannot be expected to stabilize the economy over a longer period, let us say of 50 years, in the presence of the changes occurring in the world economic system.

This conclusion may seem to sound the death knell for the Euro: if the measures presently envisaged – the definition and acceptance of which are very difficult, as demonstrated by the slow and laborious negotiations amongst member states – are not enough, what hope can there be for the survival of the euro? To answer this question we have to bear in mind that the creation of the EU was one of the most important institutional innovations of the second half of the twentieth century and that the process of constructing an adequate institutional structure for the EU is still underway.

The diverging competitiveness of the member countries

In a recent paper, De Grauwe (2010) analyses the impact of the financial crisis on the future of the Euro area and raises the question of whether the Euro area is sustainable in the long run. He concludes that the survival of the Euro area hinges on the capacity of its leaders to improve the Euro area's governance. In particular, he observes that the present crisis has mostly to do with divergent developments in the private debt of member countries, which are related to macroeconomic divergences in general. In explanation of these residual macroeconomic divergences he proposes the business models of different member countries which originate from what he calls their *animal spirits*, which he defines as waves of optimism and pessimism that, in a self-fulfilling way, drive economic activity (see Akerlof and Shiller, 2009; Leijonhufvud, 1973; Minsky, 1986). Such animal spirits differ amongst member countries and constitute an obstacle to economic integration. Examples of such animal spirits are the "angst" that prevailed in Germany a few years ago while bursts of optimism exploded in Spain and Ireland. Thus, member states of the Euro area still behave as independent nations creating their own animal spirits.

It is to be observed that devaluation seemed to be a part of the business model of southern, but not of northern, Euro area countries, before they joined the Euro area (De Grouwe, 2010). However, devaluation is not the only possible way to restore competitiveness. Devaluation reduces production costs and makes a country competitive only in cost terms. As a country becomes richer, devaluation tends to lose effectiveness as the only or the dominant way to increase competitiveness. While the need to reduce costs can never be neglected, if used alone devaluation becomes self-defeating and has to be combined with increasing knowledge and innovation intensity. A country can more easily tolerate a high exchange rate if there are some fields of industry and technology in which it has a temporary monopoly. Thus, the loss of the freedom to use devaluation only anticipated or made more explicit the need for southern Euro area countries to increase their knowledge intensity and innovation capabilities.

Animal spirits and innovation systems

The existence of these idiosyncratic business models (i.e. animal spirits) is at the core of the divergences in competitiveness observed during the last decade. The author of this chapter shares the awareness that EU and Euro area member countries differ in substantial ways which do not seem to provide a uniform response to common environmental pressures, but seem rooted in historical and idiosyncratic influences. However, rather than attributing these differences to animal spirits, the present author prefers to attribute them to different National Innovation Systems (NIS). The concept of the NIS has been introduced to emphasize the fact that innovations occur in a system, thus involving both some components of the system and their interactions (Freeman, 1987; Lundvall, 1988, 2007; Edquist, 1997, 2005; Nelson, 1992). The NIS literature stresses the point that changes in the relative performance of different countries in the world economic system does not only (or mainly) depend upon patterns of resource allocation at a given time, but also upon the institutions which interact with and contribute to scientific production, technological innovation and industrial performance. In particular, the NIS concept seems to account for the observed persistent asymmetry of the output structure and of the institutional configurations of individual countries. Even though the constraints faced by different countries may be common, no two countries have the same output structure or institutional configuration. However, this does not mean that countries cannot change. In fact, after World War II countries such as Japan, South Korea and China underwent a remarkable development processes.

The most typical components of an NIS are education, research and industrial production. However, the boundaries of an NIS are not rigidly defined and we can distinguish a narrow view of the NIS, which would include only the above components, and a broader view which would also include other institutions, such as the labour market and the unemployment and social security systems. In other words, a broad conception of the NIS can overlap with the social model of a country. Whether one accepts the narrow or the broader view of the NIS, it is quite clear that the labour market and the unemployment and training systems can affect the performance of an NIS.

NISs differ amongst countries, but they also tend to change in a systematic way during the process of economic development. Typically countries at very low levels of economic development rely upon imitation and low labour costs in order to start catching up. At this level they carry out very little R&D. However, progress towards the economic frontier becomes increasingly difficult as the GDP per capita of the country increases, and so to keep progressing developing countries need to create a well-functioning NIS (Fagerberg and Verspagen, 2007). A close correlation has been found between the rate of growth of countries and the quality of their NIS (Fagerberg and Shrolec, 2008). Considering that all EU and Euro area countries have wages and labour costs which are high by world standards, one would expect them to have made efforts to strengthen their NIS so as to adapt it to a world in which emerging countries are rapidly challenging developed countries (DCs) in fields of output and export which used to be their preserve. The world economic system is now undergoing a transition to a situation in which a number of presently emerging countries will have caught up with the DCs. This transition, and the final state it is likely to lead to, entail both opportunities and dangers for the present DCs. Exploiting these opportunities involves adapting the country NIS to achieve fields of temporary monopoly which can allow present DCs to adjust their wage and cost differentials to avoid the collapse of their social models and welfare states. An obvious component of such an adaptation strategy consists of moving towards more knowledge-intensive economic activities.

In spite of the above we find that while some EU and Euro area countries have considerably strengthened their NIS, other countries have not progressed at all in this direction (Tilford and Whyte, 2010). Typically we find that "wise" northern EU countries have improved their NIS, while southern countries did not do this. Thus, the diverging competitiveness between northern and southern EU and Euro area countries is likely to depend on animal spirits only to the extent that such animal spirits lead to different choices of NIS. The question then arises of why southern Euro area countries decided not to invest enough in higher education and R&D in spite of the growing evidence that they are important co-determinants of growth. The answer is that these countries did not perceive the above adaptation strategy as the only possible one, but instead continued to rely upon a combination of low labour costs and devaluation as long as possible. To keep pursuing such a strategy became impossible after joining the Euro area. It needs to be observed that such a strategy would have become inefficient even in the absence of the euro as their economies became richer and unable to compete purely on a cost basis. By persisting with a low R&D and higher education strategy these countries substantially undermined their competitiveness. Thus, the increasing divergence in competitiveness and the macroeconomic crises have their roots, at least in part, in different conceptions and realization of NISs.

One of the main challenges in defining a plan to overcome the present Euro area crisis consists of reinforcing the NISs of the member countries. The overarching objective of these should be the creation of a "learning economy" (Lundvall, 2002). A learning economy is an economic system in which learning activities are continuously carried out. The need for such an economic system comes from the growing rate of creation and from the growing differentiation of new knowledge. More than half of human knowledge has been created since the 1960s, and the amount of new knowledge increases exponentially (De Solla Price, 1963). Along with a growing amount of new knowledge goes an increasing differentiation of both material and knowledge production, which is the outcome of the process of structural change which accompanies economic development. Structural change is a change in the composition of the economic system which follows from the emergence of new economic activities and new sectors and the decline and extinction of older ones. Structural change is inherent in the evolution of technologies. As industrial sectors mature their capital intensity tends to grow and their labour intensity tends to fall. Furthermore, a trend which has become evident since the early twentieth century is the growing quality and differentiation of most products and services. The trends described above can be summarized by the following three long-run trajectories of economic development (Saviotti and Pyka, 2012):

Trajectory 1: The efficiency of productive processes increases during the course of economic development. Here efficiency must be understood as the ratio of the inputs used to the output produced, when the type of output remains *constant*.

Trajectory 2: The output variety of the economic system increases in the course of time. Here such variety is measured by the number of distinguishable sectors, where a sector is defined as the set of firms producing a common although highly differentiated output.

Trajectory 3: The output quality and internal differentiation of existing sectors increases in the course of time after their creation. This means that if the type of output changes during the period of observation, what we will observe is a combination of growing productive efficiency and of quality change.

From now on we will use the term variety as a synonym for diversity, although the two are in principle not identical. Such variety can exist at the inter-sectoral level as well as at the intra-sectoral level. Thus, two sectors will produce completely different types of output while one sector will produce a diversified output. In the literature these two types are often described as vertical differentiation and horizontal differentiation respectively. Such long-run trajectories do not emerge separately, but exist due to a complex pattern of interactions within the economic system.

The implications of these trajectories for this chapter are as follows:

- Given their high labour costs, most EU/Euro area countries cannot expect to be competitive purely on a cost basis. They need to compete in subsectors or product ranges which are of high quality or unique in order to create the temporary monopoly which could allow EU/Euro area countries to organize a reasonable transition towards a world in which most countries are developed. Reasonable here means a transition which allows EU/Euro area countries to preserve the substance of their social models and to avoid extreme social tensions.
- The above three trajectories are not independent. Productive efficiency needs to grow in order to generate the resources required to create higher

variety and quality. However, historically the growth of productive efficiency alone could not have given rise to the development path that we observe. Given this interdependence efficiency, variety and quality must be increased in coordination.

- Given the different levels of economic development of EU/Euro area countries one cannot expect all of them to adopt the same pattern of specialization. In particular, a number of recent studies (Frenken *et al.*, 2007; Hidalgo *et al.*, 2007; Saviotti and Frenken 2008; Boschma and Iammarino, 2007) point out that the most successful development paths involve a gradual process of raising output variety and quality rather than trying to jump suddenly from low technologies to very high ones. In other words, we cannot expect all EU/Euro area countries to suddenly create Silicon Valleys to solve their competitiveness problems. Countries with NISs of lower technological and knowledge intensity should start by gradually improving the variety, quality and knowledge intensity of their present activities while simultaneously preparing the ground for future and more radical changes in their production capability. The concept of Smart Specialization (Foray *et al.*, 2009, 2011; McCann and Ortega-Argilés, 2010) focuses on the need to start improving present activities rather than trying to immediately build high-tech forms and sectors.

The trend towards falling labour intensity which we could consider internal to the evolution of technologies is accelerated by the entry of emerging countries. Such structural change is not an epiphenomenon of economic development, but it is increasingly recognized as one of its most important determinants (see for example Murphy Shleifer *et al.*, 1989; Foellmi and Zweimuller, 2006; Matsuyama, 2002; Saviotti and Pyka 2004, 2008, 2012; De Benedictis *et al.*, 2009).

The presence of structural change implies that the composition of economic activities and the competencies required to carry them out needs to change over the course of time. When structural change was very slow we could expect the replacement of competencies to occur with the change in the generations of the labour force. However, both trends in technology and the entry of emerging countries have substantially accelerated structural change. Thus, while we could expect that sectors such as steel, shipbuilding and car production would become gradually less labour-intensive, the entry of countries such as Japan, South Korea and, more recently, China, has accelerated this trend.

The growing speed of structural change implies that the competencies of the labour force may need to be changed more often than in the past. Any attempt to keep these competencies constant is likely to backfire. The only serious solution to the problems faced by mature industrialized countries is a program of continuous learning by the labour force so as to create competencies adapted to the international economic environment. Under these conditions the objective of learning programs is to create competencies which allow EU countries to justify the relatively high wages they have in comparison with those of workers in those emerging countries which are increasingly becoming competitors. In order to achieve this objective activities in which EU countries can have a temporary monopoly have to be created. These activities need to be knowledge intensive, but not necessarily high-tech. This would allow the EU to adapt well to the transition towards a world in which most countries will be developed and will have similar wages.

The types of economic activities which could give EU countries a competitive advantage consist either of (i) producing unique products or services or (ii) producing products or services of a higher quality than those of lower labour-cost competitors. Naturally high-tech activities satisfy these requirements, but they cannot be the only solution. In a knowledge-based society all economic activities, including mature and traditional ones, need to become increasingly knowledge-intensive.

Any such strategy needs to have or to develop:

- A good education system, including both general education at all levels and technical training;
- A good research system with adequate interfaces between research organizations and industrial firms;
- An institutional framework favouring the flexible adaptation of competencies to the present situation of the labour market.

Policies

The main policy objective of the EU and the Euro area should be to plan a transition to a world situation in which most countries will be developed, starting from the present situation in which the EU and the Euro area have wages which are very high relative to those of emerging countries. This objective can only be achieved in a reasonable way, avoiding excessive social tensions by creating areas of temporary monopoly which can justify the presence of these high wages in the short run and allow a gradual realignment with the wages of emerging countries in the longer run. This path cannot be achieved only by measures aimed at increasing productive efficiency, but requires creativity. In other words, we cannot expect to solve the transition problem by reducing EU wages to the level of those of emerging countries or to reduce the labour intensity of existing industrial processes to maintain the existing wage differentials. Both of these solutions would cause either mass poverty or mass unemployment, or both. At best they can be components of an overall adaptation strategy. What is missing from these components is the creation of activities which lead to new products and services or to the increasing quality and differentiation of existing ones. Thus, an overall strategy for the EU has to be based on a combination of growing productive efficiency, growing variety and greater sectoral quality and differentiation.

While this is common to the whole EU, the wide divergence in competitiveness within the member countries will require component strategies specific to each country. Let us first examine the common components of the strategy. A necessary requirement to construct areas of temporary monopoly is to increase the knowledge intensity of all economic activities, present and future. Furthermore, the need arises to change the composition of the labour force so as to adapt it to the present situation of the world economy. This does not imply that all of the existing sectors will maintain a constant employment or even will survive. It will be necessary to reduce employment in sectors which are subject to intense international competition by emerging countries and to combine layoffs with the training of the labour force made redundant from particular jobs. The need to undergo frequent retraining is essential to conceive the labour market as constituted by those who are employed and by those who are being retrained, except for people with special conditions. While the funding of general retraining programmes might be considered an additional burden on state budgets, where it has been applied it has reduced unemployment levels. Thus, rather than requiring new resources, well-designed retraining programmes amount to a transfer of resources from unemployment to training. To the extent that unemployment is due to misadjusted competencies, paying people to stay out of the labour market will further reduce the value of their competencies and lead to long-term unemployment. This outcome is clearly inferior to the one produced by training – that is, a labour force with up to date competencies.

A generalized retraining policy of the type referred to above would find an insuperable obstacle in high barriers to the laying-off of people. Such barriers are equivalent to prevent structural change to operate. The correct approach is to move people from sectors subject to a high intensity of competition to activities in which a temporary monopoly can be established. Once more, we need to restate that simply reducing barriers to laying-off people will not be effective and could easily worsen the situation. Lower barriers to laying-off people need to be rigidly tied to retraining programmes.

A policy of this type has already been applied in Denmark under the name of Flexicurity (See Flexicurity references), and it has been very effective in reducing general and youth unemployment to 4%. It is to be observed that while Flexicurity was introduced on an empirical basis, it can find a proper logical foundation in models of economic development stressing structural change. Of course, while the general principle of a policy of this type is common to most EU countries, the detailed nature that such policies will take in each country depends upon their existing institutions and social models.

Policies of this type involve the coordination of a number of institutions, only some of which are present in the narrow view of the NIS. For example, a reform of the labour market has been essential to implement the Flexicurity policy by reducing the barriers to laying-off people and enabling the required flexibility of the labour market. It is to be pointed out that such flexibility can be properly exploited if, in addition to lowering barriers to lay-offs, the system includes retraining. Simply laying-off people and expecting the market to provide new jobs for them with ill-adapted competencies would be quite likely to worsen the problem. The implementation of a Flexicurity type policy requires the active participation of labour unions, without whose collaboration the retraining system could not function.

Some of these reforms touch upon assumptions which are at the heart of EU social models. For example, the reduction in barriers to lay-offs could be seen as a renunciation of the right to work. However, the presence of structural change implies that job security is incompatible with employment security. Our overarching objective must be to create employment security, but we can only achieve that if we are prepared to abandon job security in favour of a more flexible pattern of employment. What must be stressed is that maintaining employment security is fundamental for the survival of EU social models and welfare states. The changes proposed would preserve the goals of the social models by changing the means of their implementation.

In this sense it must be remembered that both the EU and the Euro area include countries at very different levels of economic development. However, this heterogeneity need not be just an obstacle to the improvement of the competitiveness of "weak" EU countries. Traditionally the concept of the advantages of backwardness (Gershenkrohn, 1962) have indicated the advantages constituted by the combination of low labour costs and a series of "recipes" which other countries had already developed and experimented with. Such advantages are not automatic, but need the setting up of complementary institutions which can allow their realization. Thus, simply having lower labour costs does not guarantee successful development. The problem with "weak" EU/Euro area countries is a double disadvantage which combines low technological capabilities and relatively (to these capabilities) high labour costs. In fact, what can determine competitiveness today is not just labour costs, but the ratio of wages to competencies.

The reforms required to improve the competitiveness of these countries include improvements in both the efficiency and the creativity of their economic systems. Efficiency-increasing measures have to reduce the cost of existing processes. This is likely to entail short-term reductions in wages. Such measures are likely to encounter popular resistance. Here is an opportunity to creatively use the crisis by coupling efficiency-enhancing measures with effective development plans. If short-term sacrifices are required it will be much easier to accept them when they are part of a credible plan which can be expected to lead to a future period of greater prosperity.

Efficiency-increasing policies include measures aimed at removing barriers to competition and facilitating the creation of new firms. However, while required as a component of an overall strategy aimed at improving the competitiveness of less-developed EU/Euro area countries, such measures are unlikely to be sufficient. Efficiency-based policies need to be combined with policies aimed at improving creativity and at changing the composition of the economic system.

As previously pointed out, improvements in the education of the population and the training of the labour force are crucial for improving the competitiveness of the "weak" EU/Euro area countries. This need not mean that there is the possibility to create a large number of Silicon Valleys in Europe. The concept of Smart Specialization (see Smart Specialization references) refers to a local (regional, national) specialization which starts from the existing activities of a country/region and builds upon them by making them more knowledge-intensive. The concept finds support in a number of recent studies which show that while countries need to differentiate in order to develop, they are more likely to be successful if in the short run they differentiate towards activities which are similar to those they were previously using (Frenken *et al.*, 2007; Hidalgo *et al.*, 2007; Saviotti and Frenken, 2008; Boschma and Iammarino, 2009) than if they try to "jump" to completely different economic activities. However, in order to sustain development in the longer run, a more radical differentiation of economic activities needs to be prepared before it can start paying off (Saviotti and Frenken, 2008). It follows that a development strategy for the weak EU/Euro area countries is more likely to be successful if it starts from the present activities of those countries and proceeds by making them progressively more knowledge-intensive. However, in order to sustain their development process in the course of time these countries need to start exploring potential new activities before they become profitable.

The discussion so far in this section cannot be interpreted as implying that the only changes required are improvements in education and training, although these changes are central to an adequate development strategy for weak EU countries. These changes are unlikely to be useful unless they are coupled with changes in the labour market and in a number of organizations. As already pointed out, retraining needs to be the complement to a lowering of barriers to the laying-off of people. For example, such barriers could be lowered in exchange for an engagement by producers to contribute to a retraining fund which financially supports the cost of training. Furthermore, an institution that needs to change its behaviour is that of the labour unions. In Scandinavian countries, where such retraining programs have been successfully adopted, labour unions are heavily involved in their administration, rather than having a purely adversarial attitude.

The above considerations are just examples of policy measures which could be adopted in order to improve the competitiveness of weak EU/Euro area countries. The general principle underlying these policy measures is to move all of the economic activities of these countries towards a greater knowledge-intensity in order to create regions (in knowledge or product space) of temporary monopoly. This principle could be implemented both by increasing the knowledge-intensity of existing activities and by creating new (for them) economic activities. In this way weak EU/Euro area countries could better adapt to an international economic environment in which the intensity of competition is continuously increasing in the activities which were the basis of the economic models of the EU and the USA. Such adaptation by means of an enhanced competitiveness is required both to support EU social models and to provide a socially and economically sustainable development path for the EU.

Although the above considerations are related to the need for a long-term economic strategy for the EU, some of the choices required to overcome the present crisis could have a considerable effect upon the survival of the Euro area. In particular, the choice between rapidly re-absorbing the debts incurred in the past by member countries and starting to develop strategies to attain future competitiveness and growth is crucial. Considerable attention should be paid to designing a path which, in addition to holding the promise to reduce the debt, contains measures which can be expected to lead to recovery and growth as soon as possible. The speed with which debts can be reabsorbed does not seem to worry financial markets as much as the absence of growth perspectives.

Summary and conclusions

The world has changed considerably with respect to the one in which the EU was created. If the initial objective of the EU can be considered to have been achieved, the changes which have occurred in its external environment require a modification of its strategic objectives. The process of globalization and the increasing intensity of competition coming from emerging countries challenge the balance between competitiveness and cohesion which is central to the existence and stability of EU member countries. The new globalized world includes newly emerging countries which compete in sectors which used to be the preserve of developed countries. Given that the sectors affected by this increasing competition are those which were previously called productive sectors, and that these same sectors are complementary with respect to welfare states, the new international economic environment represents a fundamental challenge for the existence of the EU. Such a challenge is magnified by the adoption of the Euro and by the present world crisis.

So far, Euro area countries have reacted in ways which would have been unimaginable before the crisis. The measures already introduced move in the direction of a greater integration of EU countries, to an extent which was difficult to imagine before the crisis. In spite of this progress such measures are unlikely to assure the long-term stability of the EU. First, such measures are incomplete: important components such as the Eurobonds or the Banking Union are still the object of considerable controversy and progress is understandably slow. Second, such measures are essentially financial and do not provide any coordination for economic and innovation policies. This chapter has argued that that even if the financial measures were to be completed and to be successful, they could only stabilize the EU, and in particular the Euro area, in the short term. Even if successful, such financial measures would at best prevent member countries from running very high debts, but would do nothing to reduce the divergence in their competiveness. In these conditions "weak" member countries would become increasingly poorer than "strong" member countries, a situation quite likely to weaken (if not to destroy) the EU. Thus, the main objective of this chapter is to show that the longer-term stabilization of the EU and the Euro area involves a greater coordination not only of financial policies but also of economic and innovation policies.

It has been pointed out that the social models of member countries play a very important role in the stability of those countries and of the EU as a whole. At the centre of these social models are the welfare states of individual member countries. Welfare states are not always the drain on resources which critics accuse them of being. Although welfare states emerge in countries which are above a minimum threshold of economic development, and can then be considered a form of high-level consumption for rich countries, in the right conditions they can be complementary with respect to the sectors which were before called productive – essentially those sectors which contribute to exports. Thus, the complementarity of productive sectors and of welfare states underlies that of competiveness and cohesion, itself extremely central to the social models of member countries and to the objectives of the EU as a whole. In the present international economic environment it is extremely important to increase the competitiveness of EU member countries in order to maintain the right balance between competiveness and cohesion. It is not only vital that the average competiveness of member countries is increased, but also that the divergence of the competiveness of individual countries is reduced. The long-term survival of the EU depends upon the capacity to improve the competiveness of "weak" countries relative to that of the "strong" ones so that at least the former do not fall behind the latter.

Beyond the details specific to each country, such increases in competiveness cannot be obtained solely by efficiency-increasing and cost-reducing measures, but require the development of sectors, subsectors and product ranges in which competitiveness is based upon variety and quality. As a consequence, increasing competitiveness will inevitably entail an increase in the knowledge-intensity of the economies of EU member countries, or, to put it differently, an improvement in their National Innovation Systems, possibly leading to a change in their patterns of specialization. This becomes even more important for member countries of the Euro area, given the more limited range of policy options that membership of the Euro entails. While the development of high-tech sectors is one of the strategic paths which can be used to raise knowledge intensity, it is not the only one, and it might not be realistic to expect all EU countries to be able to achieve it in the short or medium run. A more realistic approach can be represented by starting to raise the knowledge-intensity of existing local economic activities, gradually differentiating them and then proceeding to create more diverse economic activities in the longer run. This strategy, which is now being discussed as "smart specialization", can find a theoretical foundation in the trajectories discussed in section 5, which involve a progressive differentiation of the economic system, and in the observation that the process of differentiation is more realistic and feasible if the differentiation is gradual and proceeds by learning initially based around existing activities and then moving towards greater differentiation.

A process of increasing knowledge-intensity and output differentiation will inevitably require a change in the competencies of the labour force of EU countries. Thus, a strategic reorientation of this type cannot be realized simply by changing the allocation of resources to R&D and to higher education, however necessary they are; it will require changes in a number of institutions, starting from the labour market, unemployment insurance and the training and retraining systems. Examples of policies combining this range of institutions already exist, the so-called Flexicurity being the most prominent one. However, policies broadly aimed in the same direction will need to be adapted to the local reality of each member country.

A strategic reorientation of this type can either be left to individual member countries or be in part coordinated at the EU level. Given the success achieved by the CEC in coordinating European research it seems desirable to provide at least some EU-level coordination for innovation and industrial restructuring. If this will certainly involve resources, it seems better to use these resources for a constructive purpose, thus reinforcing the competitiveness of all EU countries, rather than continually bailing out the weak ones when crises arise.

References

Akerlof, G. and Shiller, R. (2009), *Animal Spirits: How Human Psychology Drives the Economy and Why It Matters for Global Capitalism*, Princeton University Press, 264pp.

Boschma, R. A. and Iammarino, S. (2009), "Related Variety, Trade Linkages and Regional Growth", *Economic Geography*, 85(3), 289–311.

Clarysse, B. and Muldur, U. (2001), "Regional cohesion in Europe? An analysis of how EU public RTD support influences the techno-economic regional landscape", *Research Policy*, 30 275–296.

Competitiveness and Cohesion [No Author] (2011): www.europedia.moussis.eu/books/Book_2/5/13/02/index.tkl?all=1&pos=150

De Benedictis, L., Gallegati, M. and Tamberi, M. (2009), "Overall trade specialization and economic development: countries diversify!", *Review of World Economics*, 145(1): 37–55.

De Grauwe, P. (2010), "The Financial Crisis and the Future of the Eurozone", Bruges European Economic Policy Briefings, BEEP n° 21.

De Grauwe, P. (2011), "Managing a fragile Eurozone", Cesifo Forum, 2/2011.

De Solla Price, D. (1963), *Little Science, Big Science*, New York, Columbia University Press.

Economist, The (2012), Asian Welfare States: New Cradles to Graves, pp. 20–22, 8th September.

Economist, The (2012), Special report: China's economy, 26th May.

Economist Intelligence Unit (EIU) (2011), Special report: State of the union, Can the euro zone survive its debt crisis? (March).

Edquist, C. (1997), *Systems of Innovation: Technologies, Institutions and Organizations*, London: Pinter.

Edquist, C. (2005), "Systems of innovation, perspectives and challenges", in Fagerberg, J., Mowery, D. and Nelson, R. R., *The Oxford Handbook of Innovation*, Oxford: Oxford University Press, pp. 181–208.

Eichengreen, B. (2011), "Can the Euro Area Hit the Rewind Button?", University of California, Berkeley, 23rd July.

Esping Andersen, G. (1999), *Les Trois Mondes de l'Etat Providence*, PUF.

Fagerberg, J. and Shrolec, M. (2008), "National innovation systems, capabilities and economic development", *Research Policy*, 37, 1417–1435.

Fagerberg, J. and Verspagen, B. (2007), "Innovation, growth and economic development: have the conditions for catch-up changed", *International Journal of Technological Learning, Innovation and Development*, Vol. 1, No. 1, 13.

Flexicurity

- Algan, Y. and Cahuc, P., Civic Attitudes and the Design of Labor Market Institutions: Which Countries Can Implement the Danish Flexicurity Model? Discussion Paper No. 1928, Forschungsinstitut zur Zukunft der Arbeit Institute for the Study of Labor January 2006
- Andersen, T. M., Bosch, N., Deelen, A. and Euwals, R., The Danish flexicurity model in the Great Recession, 8 April 2011, www.voxeu.org/article/flexicurity-danish-labour-market-model-great-recession.
- European Foundation for the Improvement of Living and Working Conditions (2007) "Flexicurity", www.eurofound.eu.int/areas/industrialrelations/dictionary/definitions/FLEXICURITY.htm.
- Larsen, T. P., Hansen, J. A., Hansen, N. W., Danish Flexicurity – A Role Model, but what is there to Copy? – The Case of Fixed-term, Agency, and Visiting Workers, Paper to the IIRA Copenhagen, June 28th to July 1st 2010
- www.euractiv.com/socialeurope/flexicurity-europes-employment-s-linksdossier-188451

Foellmi, R. and Zweimuller, J. (2006), "Income distribution and demand induced innovations", *Review of Economic Studies*, 82, 95–112.

Freeman, C. (1987), *Technology Policy and Economic Performance*, London, Pinter.

Frenken, K., van Oort, F. G. and Verburg, T. (2007), "Related variety, unrelated variety and regional economic growth", *Regional Studies*, 41(5): 685–697.

Funke, M. and Ruhwedel, R. (2001a), "Product variety and economic growth: Empirical evidence for the OECD countries", *IMF Staff papers*, Vol. 48, N° 2.

Funke, M. and Ruhwedel, R. (2001b), "Export variety and export performance: empirical evidence from East Asia", *Journal of Asian Economics*, Vol. 12, 493–505.

Gerschenkron, A. (1962), *Economic Backwardness in Historical Perspective* Cambridge, Mass: Harvard University Press.

Hidalgo, C. A., Klinger, B., Barabasi, A. L. and Hausmann, R. (2007), The product space conditions the development of nations. *Science*, 317: 482–487.

Josselin, J. M. and Marciano A. (2006), The political economy of European federalism, Working Paper, CREM, Université de Rennes 1 and Centre National de la Recherche Scientifique.

Lefevre, A. and Meda, D. (2006), *Faut-il bruler le modèle social français?* Paris: Seuil.

Leijonhuvud, A. (1973), Effective demand failures. *Swedish Journal of Economics*, 75: 27–48.

Lundvall, B. A. (1992), *National Systems of Innovation*, London: Pinter.

Lundvall, B. A. (2002), *Innovation, Growth and Social Cohesion: The Danish Model*, Cheltenham, Edward Elgar.

Lundvall, B. A. (2007), "National innovation systems: from List to Freeman", in Hanusch, H. and Pyka, A. (eds), *Elgar Companion to Neo-Schumpeterian Economics*, Cheltenham, Edward Elgar.

Matsuyama, K. (2002), "The Rise of Mass Consumption Societies", *Journal of Political Economy*, 110: 1035–70.

Minsky, H. (1986), *Stabilizing an Unstable Economy*, McGraw Hill, 395pp.

Mundell, R. A. (1961), "A Theory of Optimum Currency Areas", *American Economic Review*, 51, 657–665.

Mundell, R. A. (1968), *International Economics*, New York: Macmillan.

Murphy, K., Schleifer, A. and Vishny, R. (1989), "Industrialization and the Big Push", *Journal of Political Economy*, 97, 1003–26.

Nelson, R. R. (ed) (1992), *National Innovation Systems*, Oxford, Oxford University Press.

Official Journal C 321E of 29 December 2006.

Rose, A. K. (2006), *Currency Unions*, The New Palgrave.

Sapir, A. (2005), *Globalization and the reform of the European Social Models*, Bruegel Policy Briefs, November 2005.

Saviotti, P. P. and Frenken, K. (2008), "Export variety and the economic performance of countries", *Journal of Evolutionary Economics*, Vol. 18, 201–218.

Saviotti, P. P. and Pyka, A. (2004), "Economic development by the creation of new sectors", *Journal of Evolutionary Economics*, 14(1), 1–35.

Saviotti, P. P. and Pyka, A. (2008), "Product Variety, Competition and Economic Growth", *Journal of Evolutionary Economics*, Vol. 18, 323–347.

Saviotti P. P. and Pyka, A. (2012), "On The Co-Evolution Of Innovation And Demand: Some Policy Implications", Revue de l'OFCE, n°124 – *Debates and policies: Agent-based models and economic policy*, edited by Jean-Luc Gaffard and Mauro Napoletano.

Sharp, M. (1998), "Competitiveness and cohesion – are the two compatible?", *Research Policy*, 27, 569–588.

Smart Specialization

- Foray, D., David, P. A. and Hall, B. (2009), "Smart Specialisation – The Concept", Policy Briefs delivered by the "Knowledge for Growth" Expert Group advising the Commissioner for Research, Janez Potočnik, http://ec.europa.eu/invest-in-research/monitoring/knowledge_en.htm.
- Foray, D., David, P. A. and Hall, B., (2011) "Smart specialization, From academic idea to political instrument, the surprising career of a concept and the difficulties involved in its implementation", Management Of Technology & Entrepreneurship Institute(Mtei), *Mtei-Working_Paper-2011–001.*
- http://s3platform.jrc.ec.europa.eu/home.
- McCann, P. and Ortega-Argilés, R. (2011), "Smart Specialisation, Regional Growth and Applications to EU Cohesion Policy", Economic Geography Working Paper 2011: Faculty Of Spatial Sciences, University Of Groningen

Tilford, S. and Whyte, P. (2010), "The Lisbon scorecard X: The road to 2020", Centre for European Reform (CER).

Treaties of Rome:

- www.historiasiglo20.org/europe/traroma.htm ()
  ww.cvce.eu/obj/treaty_establishing_the_european_economic_community_rome_25_march_1957-en-cca6ba28–0bf3–4ce6–8a76–6b0b3252696e.html.

4 Innovation and economic performance (industrial and financial)

Recent results and questions for future research

Pelin Demirel and Mariana Mazzucato

1 Introduction

Despite very strong differences in their treatment of technological change in economic theory, both the neoclassical and the Schumpeterian evolutionary economic approaches often assume that market selection rewards the most innovative firms. In other words, more innovative and, hence, more efficient firms should outperform the less innovative ones, achieving higher growth, profits, and stock prices. This is because product innovation can create new markets and/or increase the market shares of innovators, while process innovations improve the productivity of innovators by cutting production costs. However, despite such strong assumptions, empirical evidence on whether innovative firms perform better than non-innovative firms remains inconclusive.

While the positive effect of innovation on financial performance, such as market value and stock prices, has been found to be more or less robust (Blundell *et al.*, 1999; Griliches, 1984; Hall *et al.*, 2005; Toivanen *et al.*, 2002), the empirical evidence for the impact of innovation on firm growth is much more mixed and does not firmly confirm the assumption that innovative differences among firms lead to growth differentials. Dosi (2005) formulates this problem in the following words: "The impact of both innovativeness and production efficiency upon growth performances appears to be somewhat controversial.... Contemporary markets do not appear to be too effective selectors delivering rewards and punishments according to differential efficiencies" (pp. 25, 29).

If innovators do not grow more, does this imply that market selection fails? And does the different impact of innovation upon industrial performance (measured by firm growth and profitability) and financial performance signal differences in how industrial and financial markets react to firm-level efforts around innovation?

Industrial economists are faced with a related puzzle as one digs deeper into the time series properties of innovation and firm performance variables, as well as the relationship between these. While innovation and profits show some degree of *persistence* in their time series behaviour, firm growth lacks much persistence. How can one reconcile the dissimilar dynamics of innovation and firm growth variables if the underlying assumption of market selection is that the innovative behaviour of firms should translate to performance? Should not the persistent innovators grow more persistently? In what follows, we review the literature around these puzzles and discuss potential ways to address these issues.

2 Market selection

Friedman (1953) argued that the survival of firms depends upon their ability to maximize profits. He argued that firms which fail to maximize profits will eventually be driven out of the market. While it is not unreasonable to assume that the more profitable firms will grow at the expense of those that are less profitable, the "profit maximization" principle stands out as an insufficient means of describing what "fitness" (i.e. relative competitiveness) entails (See Nelson and Winter, 2002, for a detailed critique of Friedman and followers). Fitness is the *relative* rather than the *absolute* efficiency of firms, upon which firm growth and survival is determined. Dosi and Nelson (1994) argue that fitness is determined by a combination of several different efficiency criteria at the firm level, the *cash flow situations, accounting profits, investor expectations regarding profitability, relative quality of products, prices and after-sales servicing, delivery delays* and *marketing networks* being some of these. In this picture, innovation stands as a crucial factor that can improve the fitness of a firm through boosting several of these dimensions of firm efficiency.

The market selection processes that choose winners and losers based on their relative efficiency supposedly favour those who operate more efficient technologies. Firms operating less efficient technologies have smaller profits and, consequently, less money to fund their growth. As the industry evolves, the market shares of the inefficient firms converge to zero and they are driven out of the market (Beker, 2004). Freeman (1995) holds that the fastest growing firms have a larger capacity for a flow of incremental innovations and occasional radical innovations. Bigsten and Gebreeyesus' (2008) work finds empirical evidence showing that productive and efficient firms are more likely to grow fast and, through this process, resources are "reallocated from less to more productive firms". Shiferaw (2007), using the same data set as Bigsten and Gebreeyesus (2008), adds that firm investments into technological capabilities enhance firm-level efficiency, which in turn allows the firm to perform better than the rest of the firms.

Kiyota and Takizawa (2006) find that market selection works gradually so as to eventually drive the inefficient firms out of the market: "Gradual declines in productivity ultimately cause the exit of firms from the market, which implies the existence of the 'shadow of death'" (Kiyota and Takizawa, 2006). Griliches and Regev's (1995) and Bellone *et al.*'s (2005) results also confirm that 'continuously below the average' efficiency results in firm exit.

Of course, the degree to which the market selection is tolerant to temporary losses of *efficiency* determines the time-horizon in which inefficient firms are driven out (Bellone *et al.*, 2005). Too short a time-horizon may lead to the failure of the most innovative firms which undertake the biggest risks with a view to long-term gains. The other extreme is the slow operation of the selection mechanism that will lead to efficiency losses on the whole as the market takes too long to reallocate the resources to more productive firms. Geroski and Mazzucato (2002) show that the capital markets ease the selection pressures on innovative firms by providing them with the means to survive until their innovative products make it to the market. Given that the capital markets do their job well by financing the most innovative firms, myopic selection based upon product merit rewards innovation, rather than punishing it.

Product innovations should play a significant role in the market selection process as consumers base their purchase decisions on the current merits of the products (Geroski and Mazzucato, 2002). On the other hand, process innovations contribute to the efficiency of firms by reducing production costs. Bellone *et al.*'s (2005) results indicate that product-innovation-based efficiency applies more for small firms, while for the large firms productivity due to process innovations is the key criterion upon which market selection rewards the winners.

Yet, it is not always clear that the market selection works as smoothly as anticipated. Indeed, there are growing concerns that the selection mechanism does not quite do the job of correctly rewarding the most innovative and efficient firms in industrial markets. For instance, Tamagni (2007) shows that firms which exit the Italian markets are not always the least efficient or the slowest growing firms (see also Nishimura *et al.*, 2005, for a similar result for Japan). Bottazzi *et al.* (2002) find that market selection operates rather gently on the 'near-average' firm but "its role, it seems, is mainly to cut out the very worst performers". In the following parts of this section, we investigate the role of innovation in the operation of industrial markets and financial markets.

2.1 Market selection in industrial markets: innovation, profitability and firm growth

The empirical evidence on the impact of innovations for profits and firm growth (indicators of firm performance in industrial markets) is mostly mixed, particularly for the latter. Geroski *et al.* (1993) find a small positive impact upon profit margins for a firm's successful innovations, while Geroski and Machin (1992) point out the persistent and significant differences in profitability of innovators and non-innovators, and Freel (2000) argues that such profitability differences between the groups of firms are contingent upon several factors, such as firm size and industry characteristics. Leiponen (2000) also finds persistent differences in determinants of profitability for innovators and non-innovators. Factors such as patenting and educational competencies positively affect the profitability of innovators, while such factors have no significant (or even negative) effects upon the profits of non-innovators. In a sample of UK firms, Stoneman and Kwon (1996) find that those firms which fail to adopt new technologies experience reduced profits, while the adopters of new technologies gain an annual gross profit of 11% above the mean profit of the sample firms. While there seems to be a relationship between innovation and profitability, these studies fail to consistently establish the exact nature of the relationship. Contradictory findings also exist, such as in Robson and Bennett (2000), who find no evidence of higher profitability for a sample of innovating small UK firms.

As far as the firm growth empirical literature is concerned, we find even more mixed results. While some studies confirm that firms which are more innovative experience more growth, others fail to identify such a clear relationship. Adamou and Sasidharan (2007) find that firms which have higher R&D intensity ratios (i.e. R&D/sales) grow faster. Also, Yasuda (2005) shows that R&D expenditures per employee have a positive impact upon firm growth. Yang and Huang's (2005) work on Taiwan's electronics industry confirms that R&D is an important determinant of firm growth. Foray *et al.* (2007) argue that R&D expenditures are positively correlated with sales growth, while Del Monte and Papagni (2003) show that R&D has a positive impact upon firm growth, but that this is more pronounced in traditional industries rather than high-tech ones. Coad and Rao's (2008) findings similarly reveal the positive impact of innovative activities upon firm growth, and they find that the impact of innovation is significantly higher for the fastest growing firms. Geroski and Toker (1996) find that innovations of the 209 large UK firms in their sample have a positive impact upon annual turnover, while Geroski and Machin (1992) identify that the firms in their sample which produced at least one "major" innovation grew faster than firms that produced no such innovation. On the other hand, the literature also presents counter-evidence regarding the impact of innovation upon firm growth. Heshmati and Lööf (2006) find no significant impact upon firm growth for the R&D expenditures of 931 Swedish firms. Kirchhoff *et al.* (2002) also find no causal relationship between the increased R&D expenditures of universities and the economic growth in a region. Oliveira and Fortunato (2005) find that physical investments have a much higher impact upon boosting firm growth compared to R&D investments. More interestingly, this holds more strongly for hi-tech firms: Bottazzi *et al.* (2001) find that innovations do not determine the sales growth of the pharmaceutical firms. Almus and Nerlinger (1999) conclude that firm growth is only affected by firm size and age and not by innovative activities. Brouwer *et al.* (1993) find that innovative activities only boost the growth of above-the-average R&D spenders.

There are also instances where studies show that innovative activities have a negative impact upon firm growth (Folkeringa *et al.*, 2004). Brouwer *et al.*'s (1993) results suggest that while product-related R&D leads to positive growth, process-related R&D causes negative growth. The negative influence of innovations possibly reflects the high cost of research, which cannot always be recovered through increased sales or profits. Teece (1986) argues that there is no obvious reason why we should assume that innovations will be translated into higher revenues or market shares for the innovators. He gives several examples from firms whose innovations were very important and first to the market, but yet they could not maintain their market lead due to a lack of "complementary" assets such as marketing, distribution and business networks.

Demirel and Mazzucato (2012) investigate the relationship between R&D and sales growth in the context of the US pharmaceutical industry and explore whether markets reward innovators with a growth premium, or whether only certain types of firms are rewarded for undertaking R&D. The pharmaceutical industry is a particularly interesting industry for the study of this question since the number of really innovative new drugs (New Molecular Entities) has remained relatively low despite the exponential increase of R&D spending and patents since the 1980s. The authors focus on how different firm characteristics affect the impact of R&D upon firm growth: (i) firm size, (ii) firm patenting, and (iii) *persistent* patenting (five years in a row). Furthermore, as the characteristics of the innovation process, as well as patenting laws, have undergone fundamental changes largely attributed to advances in genomics and IT and the 1980 Bayh-Dole act (Gambardella, 1995), they test whether the changes in the "innovation regime" have affected the R&D-growth relationship over time. Their results confirm that it can be very misleading to assume that R&D spending will translate into growth for companies. They identify a general positive impact of R&D on firm growth at the industry level, but a closer look into this relationship for different types of firms reveals that the positive impact of R&D is limited to only certain types of firms. Small pharmaceutical firms are more likely to grow as a result of their R&D efforts, but this growth is conditional upon the firms patenting persistently. For large firms, on the other hand, their results suggest that R&D fails to boost growth, and even has a negative effect upon growth performance.

The mixed results from the literature on innovation and growth suggests that innovation affects the growth of *a certain subset of firms* with certain characteristics. Thus, in order to identify a clear and significant relationship between firm growth and innovation, it is fundamental for different types of firm characteristics to also be targeted.

2.2 Market selection in financial markets: innovation and financial performance

When considering the relationship between financial performance and innovation, it is fundamental to consider both directions of the relationship: the effect of financial markets (venture capital, credit ratings, stock market reactions) on firm innovation, and the effect of firm innovation on financial markets. The characteristics of finance affect the characteristics of firm investment, and the characteristics of firm investment affect firm finance. This relationship is sector-specific and in industries where innovation is driven by new entrepreneurial firms (i.e. typically Schumpeter 1 type industries), credit creation by banks and venture capital is fundamental to facilitating innovation and economic development. In industries where, instead, innovation is driven by large persistent enterprises (i.e. typically Schumpeter 2 type industries), the self-financing of innovative investment by dominant enterprises (not individuals but "teams of trained specialists") is more predominant. In both cases, different types of government investments are also fundamental.

How finance affects innovation has important implications for long-run industrial growth dynamics. While a well functioning stock market and the existence of a dynamic venture capital market are often named as institutions that are fundamental to a country's national system of innovation. Recent results from the EC-funded FINNOV project, coordinated by one of the present authors, suggest that financial markets of different types often penalise the most innovative firms due to their short-term focus, undervaluing spending geared towards the long-run (FINNOV Policy Brief, 2010). These results include: (1) VC funding is problematic for innovation given the too short time horizons (3–5 years) and the lack of funding for immature industries that need funding the most; (2) There is increasing evidence that credit ratings focus too much on financial performance and not enough on industrial performance, so that in some cases it is the most "productive" firms that have the worst credit ratings, perhaps due to their greater spending on long-run growth investments into innovation (Bottazzi *et al.*, 2009); (3) The need to please shareholders has put pressure on firms to boost their share price through stock "buybacks" which have been detrimental to R&D spending (hence long-run growth and employment) and also increased inequality (Lazonick, 2010); (4) A UK-wide survey launched by Cambridge University's Centre for Business and Research (CBR) reveals that innovative firms appear to have been hit harder than non-innovative ones by the rising cost of credit, a likely consequence of the increasing risk aversion of lenders. These results show that after the 2007–2008 financial crisis, 37% of innovators vs. 29% of non-innovators reported increases in interest rates on overdrafts, and 47% of innovators vs. 38% of non-innovators reported increases in the size of arrangement fees (Cosh *et al.*, 2009).

In the remainder of this section we focus on a selection of recent results relating to the relationship between innovation and finance.

2.3 The impact of financial markets and venture capital (VC) on innovation

Innovation is not only risky, it is truly "uncertain" in the Knightian sense (Knight, 1921). It is also extremely expensive. Both the expensive and uncertain aspects of innovation make it very challenging for entrepreneurs, or existing firms that want to enter new areas, to find a financier that is willing to take on this high level of risk and uncertainty. For this reason, the provision of risk capital funding for start-up businesses is considered fundamental for developing and supporting entrepreneurial firms. Traditional banks are often too risk-averse for this level of risk, especially when the firm in question does not have enough internal equity (Brown *et al.*, 2009). Perez (2002) shows that in the history of many technological revolutions it is *finance capital* that leads the revolution in its early phase, with production capital playing a role only after the new technology is diffused. And it also requires a bit of "craziness" on the part of the financier – the ability to look beyond the immediate hurdles and probability of failure.

The focus of VC on the high-tech sector and on innovative projects makes it, in theory, a very important determinant of innovation outputs. Kortum and Lerner's (2000) widely-cited study of industrial R&D and patenting in the US found that venture capital investment was more productive than corporate R&D in producing patents. In this study, and using analysis of a patent production function with industry-level corporate R&D and VC investments as the main factors driving successful patent applications, the authors find that VC accounted for approximately 8% of industrial innovation between 1983–1992, while 3% was the corresponding figure that total corporate R&D accounted for. In other words, VC is found to be three times more effective at stimulating patents compared to corporate R&D. However, these studies do not adequately take into account the large amounts of government spending that predate these venture capital investments, with the latter benefitting in a free-riding way (Lazonick and Tulum, 2009).

Venture capital funds tend to be concentrated in areas of high potential growth, low technological complexity and low capital investment, since the latter raises the cost significantly. Since there are so many failures in the high risk area, venture capital funds tend to have a portfolio of different investments with only the tails earning high returns, leading to a very skewed distribution. Although most venture capital funds are structured to have a life of ten years, because of the management fees and the bonuses earned for high returns, venture capital funds tend to exit much earlier than ten years in order to establish a track record and raise a follow-on fund. This creates a situation whereby venture capital funds have a bias towards investing in projects where the commercial viability is established within a three to five year period. Although realizing such quick returns is sometimes possible (e.g. Google), it is not common. In the case of an emerging sector such as biotech or green-tech, where the underlying knowledge base is still in its early exploratory phase, such a short-term bias is damaging to the scientific exploration process, which requires longer time horizons and more willingness to risk failure.

The problem has been not only limited to the lack of venture capital investment in the most-needed early stage of projects, but also the objectives in the process of venture capital investments. This has been strongly evidenced in the biotech industry, where an increasing number of researchers have criticised the model of venture capital in science, indicating that significant investor speculation has a detrimental effect upon the underlying innovation. Some researchers have critiqued the model of venture capital in the US, indicating that significant investor speculation has a detrimental effect on the underlying innovation. For example, Lazonick and Tulum (2009) find that many venture-capital-funded businesses in the biopharmaceutical sector fail to produce any product, and yet they obtain access to finance on the stock market which is returned to shareholders, including the venture capitalists. Much of the initial investment in research and development is made by public institutions, thus suggesting that the venture capital investment cycle is not necessarily supportive of the innovation process, but is in fact supportive of the commercialization of innovation.

The fact that so many venture-capital-backed biotech companies end up producing nothing, yet make millions for the venture capital firms that sell them on the public market, is indicative of the problematic nature of the role which venture capital plays in the development of science and its effect upon the growth process. Pisano (2006) claims that the stock market was never designed to deal with the governance challenges of R&D entities.

2.4 The impact of innovation on financial performance

Unlike profits and firm growth variables that reflect the *actual* firm performance in industrial markets, *expected* firm performance can be captured by observing the level of stock prices and market values as well as the fluctuations and changes to these in financial markets. Studies that look into the impact of innovation upon financial firm performance build on the efficient market hypothesis, which assumes that the prices traded in the market reflect all known information and, hence, the firms' innovative potential is captured by these prices.

The main body of literature that connects stock price *levels* to innovation is that on the relationship between market values and patents (Pakes, 1985; Griliches *et al.*, 1991). Pakes (1985) starts with the presupposition that looking at patents and stock prices is a way to better understand the relationship between inducements in order to engage in inventive activity, the relationship between inventive inputs and outputs, and the effects of those outputs. The reasoning is that if patent statistics contain information about *shifts in technological opportunities*, then they should be correlated with current changes in market value since market values are driven by expectations about future growth. Hence, the question investigated is "to what degree is the stock market valuation of a firm a good proxy for inventive output" (Pakes, 1985). To approach this, he investigates the relationship between the *number* of successful patent application of firms (*unweighted* by citations), a measure of the firm's investment in inventive activity (R&D expenditure), and an indicator of its inventive output (stock market value of the firm). He finds that unexpected changes in patents and R&D are indeed associated with large changes in the market value of a firm. Yet there is a large variance to the increases in the value of the firm that are associated with a given patent. This is most likely due to the skewed distribution of the value of patents that has been found in the innovation literature. Griliches *et al.* (1991) make use of patent citation data to account for this large variance in the value of patents (as explained below, citations are an indicator of value/contribution, as with academic publications). This study finds that while a reasonable fraction of the variance of market value can be explained by R&D spending and/or the stock of R&D, patents are informative above and beyond R&D only when *citation-weighted* patents are used (unweighted patent numbers are less significant). Using a Tobin's q equation, they find a significant relationship between citation-weighted patent stocks and the market value of firms where market value increases with citation intensity, at an increasing rate. The market premium associated with citations is found to be due mostly to the high valuation of the upper tail of cited patents (as opposed to a smoother increase in value as citation intensity increases).

Various studies which focus on the effect of innovation on the *level* of stock prices come principally from the applied industrial economics literature that models growth, innovation and stock prices over the industry life-cycle (Jovanovic and MacDonald, 1994; Greenwood and Jovanovic, 1999; Mazzucato and Semmler, 1999) and the work on market values and patents (Pakes, 1985; Griliches *et al.*, 1991; Hall *et al.*, 2005). For example, Jovanovic and MacDonald (1994) make predictions concerning the evolution of average industry stock price levels around the "shakeout" period of the industry life-cycle. Focusing on the US tyre industry, they build a model which assumes that an industry is born as a result of a basic invention and that the shakeout occurs as a result of one major refinement to that invention. They predict that just before the shakeout occurs, the average stock price will fall because the new innovation precipitates a fall in product price, which is bad news for incumbents. Building on this work, Greenwood and Jovanovic (1999) also link stock prices to innovation by developing a model in which innovation causes new capital to destroy old capital (with a lag). Since it is primarily incumbents who are initially quoted on the stock market, innovations by new start-ups cause the stock market to decline immediately as rational investors with perfect foresight predict the future damage to old capital – "competence destroying innovations", in the words of Tushman and Anderson (1986).

Innovation is an uncertain *process* and has uncertain *outcomes*. Radical innovations cause changes to the status quo, rendering knowledge in the current period a bad predictor of knowledge in the next period (Tushman and Anderson, 1986). When innovation is "radical" or "competence destroying", the economic environment undergoes fundamental changes such that current conditions are no longer useful for making expectations about the future. During such periods, those firms that are seen as probable winners or losers will experience volatility in their stock prices because stock prices are expected to reflect *expectations* about (discounted) future profits (Pástor and Veronesi, 2005). Radical innovation that causes a shake-up of market shares diminishes the power of the incumbents who have a vested interest in the status quo and current performance is not a good indicator of future performance. In such circumstances, the level of market value tends to provide less useful information and investors are more likely to be influenced by the speculation of other investors, leading to "herd effects" and the type of over-reactions emphasized by Campbell and Shiller (1988).

While the relationship between the market valuation process and innovation is generally well understood, the missing link is between the industrial economics literature on "innovation and uncertainty" and the finance literature on the "risk and the volatility" of stock prices. The studies discussed so far relate stock price dynamics to innovation mainly by linking changes in the stock price *level* to innovation, rather than linking changes in the *volatility* of stock prices to innovation. It is the volatility, and not the level, of stock prices that reflects the dynamics of risk and uncertainty, yet very few studies provide insights into the relationship between innovation and the volatility of stock prices.

One well-known study that links stock price volatility to innovation is that of Shiller (2000), where it is shown that "excess volatility" – the degree to which stock prices are more volatile than the present value of discounted future dividends (i.e. the underlying fundamentals that they are supposed to be tracking according to the efficient market model) – peaks precisely during the second and third industrial revolutions. He suggests that it is precisely in uncertain situations, such as those characterized by radical technological change, that current information about "fundamentals" (i.e. current profits, dividends, etc.) are less useful for making predictions about future market values. Shiller's (2000) study uses aggregate data. Other studies suggest that there has been no trend increase for aggregate stock price volatility (Schwert, 1989, 2002) except for during particular periods of the 1970s and 1990s. However, the increase did not persist following these periods. On the other hand, firm-specific volatility has experienced a trend increase over the last 40 years (Campbell *et al.*, 2001). Based on these insights, Mazzucato and Semmler (1999) and Mazzucato (2002, 2003) have studied the relationship between innovation and stock price volatility at the *firm level* over the *industry life-cycle* when the characteristics of innovation are changing (Gort and Klepper, 1982). These studies (focused on the auto and computer industries) find that both idiosyncratic risk and excess volatility were highest precisely during those periods in which innovation was the most radical and market shares the most unstable.

Using industry-level data, Mazzucato and Tancioni (2008) find that the relationship between innovation and stock price volatility is rather mixed, and holds more strongly for the most innovative and the least innovative industries (e.g. biotechnology and textiles). Their further investigations using firm-level data provide evidence that innovation is related to both the level and volatility of stock prices in the case of the pharmaceutical industry (Mazzucato and Tancioni, 2007). Moreover, they show that small firms are subject to higher levels of volatility, and that firms generally exhibited higher levels of volatility in the post-1985 period.

3 Complexity, and persistence, in the innovation-performance relationship

One of the most important lessons that economists have learned from the recent financial crisis is the danger of assuming a normal distribution – the "bell curve" – in their models of stock returns. The normal Gaussian distribution assumes that most events cluster around the average value, so that those events that lie three or more standard deviations – symbolised by the Greek letter sigma in statistics – from the mean are essentially ignored. Such events are deemed almost impossible, or so rare that they don't matter. But are they?

Critics of this assumption draw attention to the role of *high sigma* events, which have proved to be much less rare than the bell curve assumes – such as the sudden fall of major stock markets on "Black Monday" (19 October 1987), or the recent fall of Lehman Brothers (15 September 2008) – which took all by surprise. In fact, the tails of the actual distribution of returns appears to be much "fatter" than the normal distribution implies. A distribution with fat tails means that high sigma events are not so unusual, and have a higher probability of occurring than the normal distribution allows for.

In fact, average returns are important only in situations where the underlying dynamic is one of *reversion to the mean* – what goes too high/low goes back down/up – with no positive feedback. Positive feedback – where success leads to more success, or over-investment leads to more over-investment – will cause the types of important, and less rare, outliers that constitute fat tails. For example, if over-valued shares become even more over-valued, leading to a "bubble", this is an example of positive feedback – and something that we have seen repeated in each of the recent bubbles: the railway bubble in the 1800s, the dot.com bubble in the 1990s, and the recent housing bubble. As policy makers are banging heads about how to regulate financial markets, such insights are crucial in making sure that the policies are not founded on erroneous statistical assumptions which invalidate the models.

Schumpeterian economists – economists who put technological and organizational change at the centre of competition – have been arguing that this lesson is also fundamental for understanding innovation, and hence innovation policy. Innovation is characterized by strong positive feedback, and also by permanent heterogeneity between firms, which is difficult to reconcile with distributions that give too much importance to the "average" company, and to "reversion to the mean" dynamics.

Different types of dynamic increasing returns to scale, and the underlying path-dependency, can lead to fat-tailed distributions because they work against a reversion to the mean process which is necessary in order for a Gaussian distribution to emerge. Therefore, if high profit firms remain high profits firms in the future, it is likely that profits will exhibit a high degree of kurtosis. Fat tails indicate that firm growth events are indeed "lumpy" instead of being *normally distributed, small and independent* (Dosi, 2005, p. 11). This is much in line with Schumpeter, who argued that "innovations are not at any time distributed randomly … but they are by nature lopsided and disharmonious … [G]rowth based on technical innovation [is] more like a series of explosions than a gentle and incessant transformation" (quoted in Freeman, 1995). We explore this subject further in the following section.

3.1 Persistence

To better understand the relationship between innovation and firm performance measures, one needs a good understanding of the time series behaviour of these variables. The literature finds some degree of persistence in the innovative activities of firms and their profits. Persistence of profits or innovations refers to the degree to which innovating firms at any given time will continue to innovate and firms with above the average profits will continue to make higher profits in the following periods. Innovative persistence and persistent profitability have important implications for firm growth as they would trigger dynamic increasing returns which result in lumpy (i.e. not in small, identical and incremental steps as outlined in stochastic firm growth models) and persistent (i.e. correlated) growth for the innovators and for firms with above the average profits (Cefis, 2003; Malerba *et al.*, 1997). Hence, if there is persistence in innovations and firm profits, one would expect these to show up as persistence in firm growth differentials. Yet, the firm growth studies consistently report a lack of persistence in firm growth behaviour, which is most puzzling for industrial economists.

There is generally consensus that innovation persists over time at the firm level, even though this persistence tends to weaken when one inspects innovation over longer periods of time (e.g. more than five years). Geroski *et al.* (1997) find that very few firms are persistent innovators. They show this finding is robust to how one measures innovation: either with patent data or the count of firms' *major innovations*. Cefis and Orsenigo (2001) and Cefis (2003) confirm the finding of Geroski *et al.* (1997) that very few firms are truly persistent innovators. Yet, they argue, these firms account for the majority of innovations in most sectors and, hence, there is persistence in innovative activities. Moreover, they argue, non-innovators are also persistent, such that their probability of starting to innovate is very low if they have never innovated in the past. Peters' (2006) results show very high levels of persistence in the innovation activities of both service and manufacturing firms and for both the innovator and non-innovator firms. Alfranca *et al.* (2002) also conclude that there is a significant degree of persistence in the innovation activities of food and beverage companies, and old innovators are the most likely candidates to introduce new innovations. Raymond *et al.* (2006) do not find any evidence for persistent innovation activities among the Dutch manufacturing firms.

Persistence in innovative activities results from a range of factors. The first and most important is the path-dependent nature of organizational routines in which the firm-specific innovative behaviour and capabilities are rooted (Nelson and Winter, 1982). Nelson and Winter (2002) argue that organizational routines provide a basis for behavioural continuity due to a number of reasons, such as irrational resistance to change, the high costs of learning and relearning the routines, and the possibility of organizational conflict that would arise as a result of major changes in routines. Hence, the firm-specific innovative capabilities which emerge from firm-specific routines show a large degree of continuity and persistence.

It is difficult for less innovative firms to effectively and fully imitate the organizational routines of very innovative firms due to the tacit nature of these routines. Tacitness hinders one from clearly observing the underlying structures of routines, and hence learning and imitation are not easy. Moreover, changing certain routines related to innovation would also require major changes in other routines throughout the organization due to the complex interdependence of those routines (Nelson and Winter, 1982; Teece *et al.*, 1997). As a result,

"Innovation intrinsically 'characterizes' a firm in that it creates a structural difference between innovating and non-innovating firms" (Cefis and Ciccarelli, 2005, p. 43).

The second reason for the persistence of innovations is that the sunk costs associated with building an R&D facility encourage the firm to continuously invest in R&D (Sutton, 1991,). This persistent R&D expenditure behaviour leads to persistent innovations. Third, firms often undertake multiple R&D projects. Hence, even when a single project delivers results/innovations at irregular intervals, the sum of all R&D projects is likely to deliver outputs at a regular interval (i.e. yearly), leading to persistent innovations at the firm level. Fourth, the positive feedback between innovation and profits leads to persistence in innovations. Firms that earn high profits due to their innovations can reinvest these earnings back into their innovative activities, which will likely deliver new innovations (Cefis and Ciccarelli, 2005). Finally, firms that have invested in innovations in the past are more likely to innovate in the future as a result of the positive feedback effects resulting from absorptive capacity (Cohen and Levinthal, 1989, 1990).

Innovative persistence needs to be understood in the context of industry evolution and the changing characteristics of the innovation regime. In the early stages of the industry life cycle innovations are more radical and innovative activities are distributed across a wide population of firms, while in the later stages of the industry life cycle the rate and magnitude of innovation tapers off, innovations become more cumulative, and the innovators are mostly large firms (Abernathy and Utterback, 1978; Agarwal and Gort, 1996; Gort and Klepper, 1982; Klepper, 1996, 1997; Londregan, 1990; Malerba, 2007; Mazzucato, 1998, 2002). The earlier stages of the industry life cycle are characterized by less persistence in innovations, while the level of persistence increases in the later stages of the industry evolution with the dominance of large firms. The innovative persistence literature documents that large firms tend to be more persistent in their innovative activities compared to small firms (Cefis, 2003; Cefis and Orsenigo, 2001; Rogers, 2004), with the exception of Japan where small firms are found to be more persistent innovators (Cefis and Orsenigo, 2001). Duguet and Monjon (2004) find that small and large persistent innovators are very different in terms of the origins of persistence: small persistent innovators rely on operational *learning by doing*, while large persistent innovators rely on *formal R&D*.

Firm profitability literature similarly suggests that there is a strong tendency for profits to persist. The empirical debate on the degree to which profits persist at the firm level starts with Mueller's (1977) study, which shows a high level of persistence in the profits of the American manufacturing firms. Jacobsen's (1988) study of medium and large American firms also reveals a degree of persistence in profits, even though the author interprets this as a "disequilibrium process" on the way to convergence of profits. Goddard *et al.* (2004), on the other hand, find that profit rates are rather stationary for a sample of 335 UK firms, and there is no evidence of converging to a common rate in the sample. Cable and Jackson (2008) show that 60% of the firms in their sample display non-eroding long-run profit persistence. Similarly, Odagiri and Yamawaki (1986) and Odagiri and Maruyama (2002) rule out the possibility of convergence and claim that profits are highly persistent in the case of Japan. In a comparison of large European firms, Geroski and Jacquemin (1988) find that profits above and below the norm tend to persist, especially for UK firms, whereas the profitability patterns in Germany and France show less persistence and more unpredictability. Cubbin and Geroski (1987) relate the persistence of profits to firm-level heterogeneity, while Glen *et al.* (2001) emphasize the relevance of geographical boundaries in determining the persistence of profits as they find profits to persist less in emerging economies in comparison with advanced economies. Gschwandtner (2005) finds that profit persistence does not only occur for surviving firms, but also applies to exiting firms as well.

Unlike in the case of profits and innovations, with the exception of some studies (Abbring and Campbell, 2003; Chesher, 1979; Bottazzi *et al.*, 2005; Contini and Revelli, 1989) most of the firm growth literature finds almost no persistence structure in firm growth (Chan *et al.*, 2003; Dunne and Hughes, 1994; Geroski, 1998; Hart and Oulton, 1996; Reichstein and Dahl, 2004; Sing and Whittington, 1975). Even in cases where growth is found to be persistent, the level of persistence is much lower than what one would expect in order to conclude that success breeds success and failure breeds failure. How can one reconcile the evidence for dynamic increasing returns – due to many reasons, but most importantly due to persistence in innovations – and profit rates with evidence for no persistence in firm growth rates?

One would also expect to find persistent firm growth due to the heterogeneity of firm characteristics and activities. The R&D spending levels of firms, technical knowledge, success in adopting and using innovations, productivity levels, the input combinations and the products, as well as the firm-specific strategies in investing, pricing, R&D, etc., are widely diverse even within very narrowly defined markets (Silverberg *et al.*, 1988, pp. 1033–1034). More importantly, these differentials do not disappear with the passage of time – they persist. Hence, firms are far from converging to a "representative firm" as firm specific organizational routines and the path-dependent development of capabilities ensure that the diversity is maintained (Nelson and Winter, 1982). Why, then, do these persistent firm-level differences not lead to persistent firm growth differentials? For instance, if more successful innovators grow faster than others, should not growth persist because the innovative characteristics of firms persist?

3.2 *Fat tails*

Firm growth rates do not follow a Gaussian distribution (as would be produced by a stochastic growth process), but display tent-like structures with significantly fatter (or heavier) tails compared to the Gaussian distribution (Axtell, 2001; Bottazzi and Secchi, 2005; Coad and Rao, 2008; Dosi, 2005; Reichstein and Jensen, 2005). Fat tails imply that extreme growth events happen more frequently, and the Pareto and Subbotin family distributions seem to fit the growth data much better than the Gaussian distribution. Evidence for fat tails is often interpreted as evidence for lumpy growth which results from economies of scale and scope, clustering of technological innovations and the increasing returns due to network externalities, knowledge accumulation, innovation activities and the self-reinforcing effects of the creation of managerial talents (Bottazzi and Secchi, 2005, p. 19; Dosi, 2005).

Yet, the existing literature does not actually test the relationship between fat tails and structural dynamics. It is suspected that innovation is a likely factor that would cause the suggested dynamics to emerge and drive the lumpy growth. The *disharmonious and lopsided* nature of innovations that often come in waves would imply that the growth opportunities in a market are not identical, some being extremely large compared to the rest (Freeman, 1995; Schumpeter, 1934). The heterogeneous nature of growth opportunities results in some firms (e.g. successful and persistent innovators) experiencing extreme growth because the market opportunities they capture are extraordinarily large. This thesis contributes to the literature by exploring whether innovation dynamics can really be related to the emergence of fat tails.

Also, a limited body of literature has shown that persistence in firm growth may be present for only a certain subset of firms (e.g. large firms) and only for certain industries (Contini and Revelli, 1989).

The literature has provided some answers to this persistence puzzle: why do we observe fat tails in firm growth rate distributions but no persistence in firm growth? For instance, Coad (2006) argues that the fat tails emerge due to the indivisibility of the lumpy resources used for growth instead of the factors suggested by Bottazzi and Secchi (2005) that would lead to increasing dynamic returns and auto-correlated firm growth. Hart (2000) believes that the lack of persistence in firm growth is because systematic factors such as R&D and capital investments fail to have a long-lasting persistent effect even though they have a one-off impact on firm growth. Oliveira and Fortunato (2005), very similarly, conclude that R&D and physical investments are important determinants of firm growth; yet, success resulting from these does not persist.

The majority of firms in an industry experience *very close to average* growth rates, while only a few experience big spurts of growth due to higher levels of available opportunities (Bottazzi and Secchi, 2005). Demirel and Mazzucato (2011) find that, in fact, those firms that are responsible for creating the fat-tailed distribution are precisely those that grow from their innovation spending – e.g. persistent innovators. While this study was focused on the pharmaceutical industry, it suggests that an important area for future research is to test whether in different sectors, non-average behaviour emerges due to similar reasons.

4 Future directions

New research must focus on understanding the time series behaviour of firm performance variables and the impact of innovation on firm performance at a finer level, for *specific types of firms*. Moreover, it is essential to compare how innovation affects firm performance in different types of industries (e.g. with different R&D intensity or a different degree of competition) and during different stages of the industry life cycle.

While the innovation–firm performance literature treats the issues of market selection and persistence at a more general level, with little reference to differences in firm characteristics, some firm-level studies acknowledge that the impact of innovation is indeed different for different types of firms e.g. slow growing/fast growing, above-average R&D intensity/below-average R&D intensity, low-tech/high-tech, etc. (Brouwer *et al.*, 1993; Coad and Rao, 2008, Del Monte and Papagni, 2003), and persistent vs. non-persistent patenters (Demirel and Mazzucato, 2012).

Thus, in this line it is important to uncover across a variety of sectors, and across the industry life-cycle, which firm-level characteristics are necessary for firms to reap benefits from their innovative activities. This agenda is of the utmost relevance to innovation and competition policies. The European Commission's EC 2020 agenda advocates that firms should spend more on R&D with a target of 3% of GDP in aggregate figures. But, if R&D does not necessarily boost firm growth, as suggested by the mixed empirical evidence, and if innovative firms do not outperform non-innovative firms, it may be potentially dangerous to ask firms to spend more on R&D. Studies of this kind will allow such policies to be better targeted towards specific firms, and also ensure that firms are developing the correct complementary characteristics to allow their R&D spending to translate into higher growth.

It is also fundamental to disentangle seemingly more straightforward relationships, such as that between innovation and market value. Some recent studies highlight that there are differences in how financial markets value innovation *in different industries* and *for different types of firms*. For example, Hall *et al.* (2005) find differences in the market valuation of innovation between sectors: elasticity tests show that the marginal effect of an additional citation per patent on market value is especially high in R&D intensive industries such as the pharmaceutical industry. Similarly, Hall and Oriani (2006) highlight the variation in how innovation affects market value across different types of firms: their findings suggest that the market only values R&D in *certain types* of Italian and French firms, specifically those in which no single shareholder holds more than one-third of the firm.

Some firm characteristics (among many) that would potentially modify the innovation–firm performance relationship include firm size (small/medium/large), patenting behaviour (patentees/non-patentees), persistence in innovation activity, tendency of firms to self-cite their patterns (an indicator of the degree to which a firm builds upon its own knowledge, generality and originality of the patenting activity – i.e. a measure based on the qualitative differences in citations made and received – Trajtenberg *et al.*, 2002), degree of involvement in inter-firm research collaborations, and firms' R&D and advertisement spending patterns (above/below the industry average and persistence in spending behaviour).

It is also important to explore the degree to which one can generalize the operating principles of market selection across industries. For example, is innovative persistence as important in terms of how market selection rewards innovators in different industries with different characteristics? A useful comparison would be cross-checking the operations of market selection across Pavitt's sectoral taxonomy in order to identify which firm characteristics underlie the market selection dynamics in different sectors. Similarly, one needs to identify whether the operating principles of market selection change as an industry progresses through its life cycle to see whether a firm characteristic (e.g. size, R&D collaborations) becomes more important at a certain stage of the industry life cycle.

It is important to understand the degree to which divergences from normality in the distribution of performance variables are related to firm-specific innovation characteristics and industry-specific dynamics. Rather than criticising the obsession of economists with normal distributions in a general way, it is important to better understand whether non-Gaussian characteristics are created by certain firms, only in certain periods of the industry history, and just for some industries. This will bring us closer to Malerba's (2006) call for evolutionary economics:

> The challenge for [coevolutionary] research here is to go to a much finer analysis at both empirical and theoretical levels, and to move from the statement that everything is coevolving with everything else to the identification of what is coevolving with what, how intense is this process and whether indeed there is a bi-direction of causality.
>
> (Malerba, 2006: p. 18)

Acknowledgements

The research leading to these results has received funding from the European Community's Seventh Framework Programme (FP7/2007–2013) under grant agreement n° 217466.

References

Abbring, J. H. and Campbell, J. R. (2003). A Structural Empirical Model of Firm Growth, Learning, and Survival. IZA Discussion Paper Series IZA DP No. 781.

Abernathy, J. and Utterback, J. M. (1978). Patterns of Industrial Innovation. *Technology Review, 88*.

Adamou, A. and Sasidharan, S. (2007). *The Impact of R&D and FDI on Firm Growth in Emerging-Developing countries: Evidence from Indian Manufacturing Industries*, from SSRN: http://ssrn.com/abstract=987024.

Agarwal, R. and Gort, M. (1996). The Evolution of Markets and Entry, Exit and Survival of Firms. *The Review of Economics and Statistics,* 78(3): 489–498.

Alfranca, O., Rama, R. and Tunzelmann, N. v. (2002). A Patent Analysis of Global Food and Beverage Firms: The Persistence of Innovation. *Agribusiness, Vol. 18* (3), 349–368.

Almus, M. and Nerlinger, E. A. (1999). Growth of New Technology-Based Firms: Which Factors Matter? *Small Business Economics 13*, 141–154.

Axtell, R. L. (2001). Zipf Distribution of US Firm Sizes. *Science 293(5536)*, 1818–1820.

Beker, P. F. (2004). Are Inefficient Entrepreneurs Driven Out of the Market? *Journal of Economic Theory, 114*, 329–344.

Bellone, F., Musso, P., Nesta, L. and Quéré, M. (2005). Productivity and Market Selection of French Manufacturing Firms in the Nineties. *WP-GREDEG-2005–6*.

Bigsten, A. and Gebreeyesus, M. (2008). Firm Turnover and Productivity Differentials in Ethiopian Manufacturing, *Journal of Productivity Analysis*, *29*(2), 113–129.

Blundell, R., Griffith, R. and Van Reenen, J. (1999). Market Share, Market Value and Innovation in a Panel of British Manufacturing Firms, *The Review of Economic Studies*, *66*(3), 529–554.

Bottazzi, G., Cefis, E. and Dosi, G. (2002). Corporate Growth and Industrial Structures: Some Evidence From the Italian Manufacturing Industry. *Industrial and Corporate Change, 11*(4), 705–723

Bottazzi, G., Dosi, G., Lippi, M., Pammolli, F. and Riccaboni, M. (2001). Innovation and Corporate Growth in the Evolution of the Drug Industry, *International Journal of Industrial Organization 19*, 1161–1187.

Bottazzi, G., Grazzi, M., Secchi, A. and Tamagni, F. (2009). Financial and Economic Determinants of Firm Default, FINNOV Discussion Paper (4.1, *in progress*).

Bottazzi, G. and Secchi, A. (2005). Growth and Diversification Patterns of the Worldwide Pharmaceutical Industry. *Review of Industrial Organization*, *26*(2), 195–216.

Brouwer, E., Kleinknecht, A. and Reijen, J. O. N. (1993). Employment Growth and Innovation at the Firm Level. *Journal of Evolutionary Economics, 3*(2), 153–159.

Brown, J., Fazzari, S. and Petersen, B. (2009). Financing Innovation and Growth: Cash Flow, External Equity, and the 1990s R&D Boom, *Journal of Finance,* American Finance Association, vol. 64(1), 151–185.

Cable, J. R. and Jackson, R. H. G. (2008). The Persistence of Profits in the Long Run: A New Approach. *International Journal of the Economics of Business,* 15, 229–244.

Campbell, J., Lettau, M., Malkiel, B. G. and Xu, Y. (2001). Have Individual Stocks Become More Volatile? An Empirical Exploration of Idiosyncratic Risk, *The Journal of Finance*, *56*(1), 1–43.

Campbell, J. and Shiller, R. (1988). Stock Prices, Earnings, and Expected Dividends, *The Journal of Finance*, *43* (3), 661–667.

Cefis, E. (2003). Is There Persistence in Innovative Activities? *International Journal of Industrial Organization, 21*, 489–515.

Cefis, E. and Ciccarelli, M. (2005). Profit Differentials and Innovation. *Economics of Innovation and New Technnology Vol. 14*(1–2), 43–61.

Cefis, E. and Orsenigo, L. (2001). The Persistence of Innovative Activities A Cross-countries and Cross-sectors Comparative Analysis. *Research Policy, 30*, 1139–1158.

Chan, L. K. C., Karceski, J. and Lakonishok, J. (2003). The Level and Persistence of Growth Rates. *Journal of Finance*, *58*, 643–684.

Chesher, A. (1979). Testing the Law of Proportionate Effect. *The Journal of Industrial Economics*, *27*(4), 403–411.

Coad, A. (2006) A Closer Look at Serial Growth Rate Correlation. LEM working paper series.

Coad, A. and Rao, R. (2008). Innovation and Firm Growth in High-tech Sectors: A Quantile Regression Approach. *Research Policy*, *37*(4), 633–648.

Cohen, W. M., and Levinthal, D. A. (1989). Innovation and Learning: The Two Faces of R&D, *The Economic Journal, 99*(397), 569–597.

Cohen, W. M., and Levinthal, D. A. (1990). Absorptive Capacity: A New Perspective on Learning and Innovation, *Administrative Science Quarterly, 35*, 128–152.

Contini, B. and Revelli, R. (1989). The Relationship Between Firm Growth and Labor Demand. *Small Business Economics, 1*(4), 309–314.

Cosh, A., Hughes, A., Bullock, A. and Milner, I. (2009) SME Finance and Innovation in the Current Economic Crisis. Centre for Business Research, University of Cambridge Working Paper (*in progress*).

Cubbin, J. and Geroski, P. (1987). The Convergence of Profits in the Long Run: Inter-Firm and Inter-Industry Comparisons, *The Journal of Industrial Economics, 35*(4), 427–442.

Del Monte, A., and Papagni, E. (2003). R&D and the Growth of Firms: Empirical Analysis of a Panel of Italian Firms, *Research Policy, 32*, 1003–1014.

Demirel, P. and Mazzucato, M. (2010). The Evolution of Firm Growth Dynamics in the US Pharmaceutical Industry, *Regional Studies*, Vol. *44*(8), 1053–1066.

Demirel, P. and Mazzucato, M. (2011). Industry Structure and Fat Tails, FINNOV working paper D2.9

Demirel, P. and Mazzucato, M. (2012). Innovation and Firm Growth: Is R&D worth it?, *Industry and Innovation*, Vol. 19, Iss. 1, 45–62.

Dosi, G. (2005). Statistical Regularities in the Evolution of Industries. A Guide Through Some Evidence and Challenges for the Theory. *LEM Working Paper Series, 2005/17*.

Dosi, G., and Nelson, R. R. (1994). An Introduction to Evolutionary Theories in Economics. *Journal of Evolutionary Economics, 4*(3), 153–172.

Duguet, E., and Monjon, S. (2004). Is Innovation Persistent at the Firm Level? An Econometric Examination Comparing the Propensity Score and Regression Methods, *University of Paris I – Cahiers de la MSE Working Paper* (Vol. No. 2004).

Dunne, P. and Hughes, A. (1994). Age, Size, Growth and Survival: UK Companies in the 1980s. *The Journal of Industrial Economics, 42*(2), 115–140.

FINNOV Policy Brief (2010, Spring). EC Framework 7 (SSH), funded project on Finance, Innovation and Growth, www.finnov-fp7.eu, 2009–2010.

Folkeringa, M., Meijaard, J. and van Stel, A. (2004). Strategic Renewal and its Effect on Small Firm Performance: Testing for Firm Size Effects Using Dutch Micro Data, *New Perspectives on Firm Growth,* 1st Inter-RENT Online Publication, 88–112.

Foray, D., Hall, B. H., and Mairesse, J. (2007, October 8–9). *Pitfalls in Estimating the Returns to Corporate R&D Using Accounting Data.* Paper presented at the First European Conference on Knowledge for Growth.

Freel, M. (2000). Strategy and Structure in Innovative Manufacturing SMEs: The Case of an English Region, *Small Business Economics, 15*(1), 27–45.

Freeman, R. B. (1995). The Large Welfare State as a System. *The American Economic Review, 85*(2), 16–21.

Friedman, M. (1953). The Methodology of Positive Economics, in *Essays in Positive Economics*, Chicago: University of Chicago Press.

Gambardella, A. (1995). *Science and Innovation: The US Pharmaceutical Industry during the 1980s*. Cambridge, Cambridge University Press.

Geroski, P. A. (1998). An Applied Econometrician's View of Large Company Performance, *Review of Industrial Organization, 13*(3), 271–294.

Geroski, P. A. and Jacquemin, A. (1988). The Persistence of Profits: A European Comparison, *The Economic Journal, 98*(391), 375–389.

Geroski, P., and Machin, S. (1992). Do Innovating Firms Outperform Non-Innovators? *Business Strategy Review, 3*(2), 79–90.

Geroski, P., Machin, S., and Van Reenen, J. (1993). The Profitability of Innovating Firms. *RAND Journal of Economics, 24*(2), 198–211.

Geroski, P. A., and Mazzucato, M. (2002). Myopic Selection. *Metroeconomica, 53*(2), 181–199

Geroski, P. A. and Toker, S. (1996). The Turnover of Market Leaders in UK Manufacturing Industry, 1979–86, *International Journal of Industrial* Organization 14(2), 141–158.

Geroski, P. A., Van Reenen, J. and Walters, C. F. (1997). How Persistently Do Firms Innovate? *Research Policy, 26*, 33–48.

Glen, J., Lee, K. and Singh, A. (2001). Persistence of Profitability and Competition in Emerging Markets. *Economics Letters, 72*(2), 247–253.

Goddard, J., Molyneux, P. and Wilson, J. (2004). The Profitability of European Banks: A Cross-sectional and Dynamic Panel Analysis. *Manchester School, 72*, 363–381.

Gort, M., and Klepper, S. (1982). Time Paths in the Diffusion of Product Innovations, *The Economic Joumal, 92*(367), 630–653.

Greenwood, J. and Boyan, J. (1999). The Information-Technology Revolution and the Stock Market, *AEA Papers and Proceedings.*

Greenwood, J, and Jovanovic, B. (1999). The IT Revolution and the Stock Market. NBER Working Paper No. 6931, available at: http://www.nber.org/papers/w6931.

Griliches, Z. (1984). Market Value, R&D, and Patents, NBER Chapters, in *R & D, Patents, and Productivity*, 249–252.

Griliches Z, Hall B. H. and Pakes A. (1991). R&D, Patents, and Market Value Revisited: Is There A Second Technological Opportunity Factor? *Economics of Innovation and New Technology, 1*(3), 183–201.

Griliches, Z., and Regev, H. (1995). Firm Productivity in Israeli Industry. *Journal of Econometrics, 65*(1), 175–203.

Gschwandtner, A. (2005). Profit Persistence in the "very" Long Run: Evidence from Survivors and Exiters, *Applied Economics, 37*(7), 1466–4283.

Hall, B. H., Jaffe, A., and Trajtenberg, M. (2005). Market Value and Patent Citations, *RAND Journal of Economics, 36*(1), 16–38.

Hall, B. H. and Oriani, R. (2006). Does the market value R&D investment by European firms? Evidence from a panel of manufacturing firms in France, Germany, and Italy, *International Journal of Industrial Organization, 24*(5), 971–993.

Hart, P. E. (2000). Theories of Firms' Growth and the Generation of Jobs, *Review of Industrial Organization, 17*(3), 229–248.

Hart, P. E. and Oulton, N. (1996). Growth and Size of Firms, *The Economic Journal, 106*(438), 1242–1252.

Heshmati, A., and Lööf, H. (2006). Investment and Performance of Firms: Correlation or Causality?, *CESIS Electronic Working Paper Series.*

Jacobsen, R. (1988). The Persistence of Abnormal Returns, Strategic Management Journal, *9*(5), 415–430.

Jovanovic, B. and MacDonald, G. M. (1994). The Life Cycle of a Competitive Industry, *The Journal of Political Economy, 102*(2), 322–347.

Kirchhoff, B. A., Armington, C., Hasan, I. and Newbert, S. (2002). The Influence of R&D Expenditures on New Firm Formation and Economic Growth. Available at *www.sba.gov/advo/research/rs222tot.pdf.*

Kiyota, K., and Takizawa, M. (2006). The Shadow of Death: Pre-exit Performance of Firms in Japan. *RIETI Discussion Paper Series 06-E-033.*

Klepper, S. (1996). Entry, Exit, Growth, and Innovation over the Product Life Cycle, *The American Economic Review, 86*(3), 562–583.

Klepper, S. (1997). Industry Life Cycles, *Industrial and Corporate Change*, *6*, 145–183.

Knight, F. H. (1921). *Risk, Uncertainty and Profit*, Boston: Houghton Mifflin.

Kortum, S. and Lerner, S. (2000). Assessing the Contribution of Venture Capital to Innovation, *RAND Journal of Economics*, *31*(4), 674–692.

Lazonick, W. (2010), The Explosion of Executive Pay and the Erosion of American Prosperity, *Entreprises et Histoire*, 57 (also FINNOV Discussion Paper 5.6).

Lazonick, W. and Tulum, O. (2009). US Biopharmaceutical Finance and the Sustainability of the Biotech Boom, Finnov Discussion Paper (5.5).

Leiponen, A. (2000). Competencies, Innovation and Profitability of Firms, *Economics of Innovation and New Technology*, 1476–8364, *9*(1), 1–24.

Londregan, J. (1990). Entry and Exit over the Industry Life Cycle. *The RAND Journal of Economics*, *21*(3), 446–458.

Malerba, F. (2006). Innovation and the Evolution of Industries. *Journal of Evolutionary Economics*, *16*(1), 3–23.

Malerba, F. (2007). Innovation and the Dynamics and Evolution of Industries: Progress and Challenges. *International Journal of Industrial Organization*, *25*(4), 675–699.

Malerba, F., Orsenigo, L. and Peretto, P. (1997). Persistence of Innovative Activities, Sectoral Patterns of Innovation and International Technological Specialization. *International Journal of Industrial Organization*, *15*, 801–826.

Mazzucato, M. (1998). A Computational Model of Economies of Scale and Market Share Instability, *Structural Change and Economic Dynamics*, *9*(1), 55–83.

Mazzucato, M. (2002). The PC Industry: New Economy or Early Life-Cycle? *Review of Economic Dynamics*, *5*(2), 318–345.

Mazzucato, M. (2003). Risk, Variety and Volatility: Growth, Innovation and Stock Prices in Early Industry Evolution, *Journal of Evolutionary Economics*, *13*(5), 491–512.

Mazzucato, M. and Semmler, W. (1999). Market Share Instability and Stock Price Volatility during the Industry Life-cycle: the US Automobile Industry, *Journal of Evolutionary Economics*, *9*(1), 67–96.

Mazzucato, M. and Tancioni, M. (2007). Stock Price Volatility and Patent Citation Dynamics: The Case of the Pharmaceutical Industry, *Open Discussion Papers* (Vol. 55).

Mazzucato, M. and Tancioni, M. (2008). Innovation and idiosyncratic risk: An Industry- and Firm-Level Analysis, *Industrial and Corporate Change*, *17*(4), 779–811.

Mazzucato, M. and Tancioni, M. (2011). R&D, Patents, and Stock Return Volatility, forthcoming *Journal of Evolutionary Economics*

Mueller, D. (1977). The Persistence of Profits above the Norm. *Economica*, *44*, 369–380.

Nelson, R. R. and Winter, S. G. (1982). *An Evolutionary Theory of Economic Change*: Bellknap Cambridge Mass.

Nelson, R. R. and Winter, S. G. (2002). Evolutionary Theorizing in Economics, *The Journal of Economic Perspectives*, *16*(2), 23–46.

Nishimura, K., Nakajima, T. and Kiyota, K. (2005). Does Natural Selection Mechanism Still Work in Severe Recessions? Examination of the Japanese Economy in the 1990s. *CIRJE-F222*.

Odagiri, H. and Maruyama, M. (2002). Does the Persistence of Profits Persist?: A Study of Company Profits in Japan, 1964–97. *International Journal of Industrial Organization*, *20*, 1513–1533.

Odagiri, H. and Yamawaki, H. (1986). A Study of Company Profit-Rate Time Series: Japan and the United States, *International Journal of Industrial Organization*, *4*, 58–66.

Oliveira, B. and Fortunato, A. (2005). Firm Growth and Persistence of Chance: Evidence from Portuguese Microdata, *ESTUDOS DO GEMF Working Papers* (Vol. 2005/10).

Pakes, A. (1985). On Patents, R & D, and the Stock Market Rate of Return, *The Journal of Political Economy*, *93*(2), 390–409.

Pástor, L. and Veronesi, P. (2005) Rational IPO Waves, *The Journal of Finance*, *60*(4), 1713–1757.

Perez, C. (2002). *Technological Revolutions and Financial Capital: The Dynamics of Bubbles and Golden Ages*. Cheltenham: Edward Elgar

Peters, B. (2006). Persistence of Innovation: Stylised Facts and Panel Data Evidence, *DRUID Working Paper Series* (Vol. 06–30).

Pisano, G. P. (2006). Can Science be a Business? Lessons from Biotech, *Harvard Business Review*, 84, no 10.

Raymond, W., Mohnen, P., Palm, F. and Van Der Loeff, S. S. (2006). Persistence of Innovation in Dutch Manufacturing: Is it Spurious?, *CEFISO Working Paper* (Vol. 1681).

Reichstein, T. and Dahl, M. S. (2004). Are Firm Growth Rates Random? Analysing Patterns and Dependencies, *International Review of Applied Economics*, *18*, 225–246.

Reichstein, T. and Jensen, M. B. (2005). Firm size and firm growth rate distributions –The case of Denmark, *Industrial and Corporate Change*, *14*(6), 1145–1166.

Robson, P. J. and Bennett, R. J. (2000). SME Growth: The Relationship with Business Advice and External Collaboration, *Small Business Economics*, *15*(3), 193–208.

Rogers, M. (2004). Networks, Firm Size and Innovation. *Small Business Economics*, *22*(2), 141–153.

Schumpeter, J. A. (1934). *The Theory of Economic Development*. Cambridge, Mass., Harvard University Press.

Shiferaw, A. (2007). Competitive Selection and Technological Capabilities in Ethiopian Manufacturing. *The European Journal of Development Research, 19*(1), 118–135.

Shiller, R. (2000). *Irrational Exuberance*. NJ, Princeton University Press.

Schwert, G. W. (1989). Why Does Stock Market Volatility Change Over Time? *The Journal of Finance, 44*(5), 1115–1153.

Schwert, G. W. (2002). Stock volatility in the new millennium: how wacky is Nasdaq? *Journal of Monetary Economics*, *49*(1), 3–26.

Silverberg, G., Dosi, G. and Orsenigo, L. (1988). Innovation, Diversity and Diffusion: A Self-Organisation Model. *Economic Journal*, *98*, 1032–1054.

Singh, A. and Whittington, G. (1975). The Size and Growth of Firms. *The Review of Economic Studies*, *42*(1), 15–26.

Stoneman, P. and Kwon, M. J. (1996). Technology Adoption and Firm Profitability, *The Economic Journal*, *106*(437), 952–962.

Sutton, J. (1991). *Sunk Cost and Market Structure*. Cambridge: MIT Press.

Tamagni, F. (2007). *An Empirical Analysis of Firm Structure and Performance: Bridging Economic Dynamics and Financial Conditions.* Unpublished PhD Dissertation, Scuola Superiore Sant'Anna.

Teece, D. J. (1986). Profiting from Technological Innovation:Implications for Integration, Collaboration, Licensing and Public Policy. *Research Policy*, *15*, 285–305.

Teece, D. J., Pisano, G. and Shuen, A. (1997). Dynamic Capabilities and Strategic Management, *Strategic Management Journal*, *18*(7), 509–533.

Toivanen, O., Stoneman, P. and Bosworth, D. (2002) Innovation and the Market Value of UK Firms, 1989–1995, *Oxford Bulletin of Economics and Statistics*, Vol. 64, 39–61.

Trajtenberg, M., Henderson, R. and Jaffe, A. (2002). University Versus Corporate

Patents: A Window on the Basicness Of Innovation. *Patents, Citations, and Innovations*, Cambridge, MA., MIT Press.

Tushman, M. and Anderson, P. (1986). Technological Discontinuities and Organizational Environments. *Administrative Science Quarterly, 31*, 439–465.

Yang, C.-H. and Huang, C.-H. (2005). R&D, Size and Firm Growth in Taiwan's Electronics Industry. *Small Business Economics, 25*, 477–487.

Yasuda, T. (2005). Firm Growth, Size, Age and Behavior in Japanese Manufacturing. *Small Business Economics, 24*, 1–15.

5 An essay on the emergence, organization and performance of financial markets

The case of the Alternative Investment Market

Valérie Revest[1] *and Alessandro Sapio*[2]

1 Introduction

Public equity represents a hardly attainable stage in the financial policy of technology-based small firms (TBSFs). The reasons are fairly clear: TBSFs enjoy rich endowments of intangible assets but lack "hard" and collateralizable assets, their track record is short, and their likelihood of survival is relatively low. The products and services offered by TBSFs largely depend on the application of scientific and technological knowledge (Allen 1992), and their founders, who often have educational backgrounds in science or technology, suffer from limited financial and marketing expertise. Partly to avoid the implied liquidity and valuation difficulties, stock exchanges usually set listing requirements in terms of minimal capitalization, profitability, free float and track record thresholds. TBSFs can rarely satisfy these conditions.

Yet, market-based financial support to TBSFs has been high on policy agendas for the last 30 years, especially in the European Union, and market segments catering to small, young, high-tech companies have been inaugurated by virtually all European stock exchanges (Bottazzi and Da Rin 2005). The apparently paradoxical attention accorded to floating companies with low survival rates stimulates reflections on the forces behind the emergence of new markets, the interests that shape their architectures, the social desirability of their performances, and ultimately on the very role of financial markets as drivers and facilitators of innovative efforts – such as those carried out by TBSFs.

This work provides an overview of the historical evolution, the organizational forms, and the performances of the stock exchanges and market segments specialized in public equity for small and growing companies, set up in Europe in the last 30 years. Our analysis mainly deals with the Alternative Investment Market (AIM), created by the London Stock Exchange (LSE) in 1995, but our discussion also involves, in passing, the "feeder" markets established in the 1970s and 1980s, as well as the "New Markets" created in the 1990s. In our view, the case under study tells a lot about four issues that are of key importance for market-based financial support to TBSFs.

First of all, let us consider the role of public powers in market emergence. Facilitating the flotation of small and growing firms can be seen as a way to solve a market incompleteness problem. This is consistent with the increasing reliance on incentive design as a key tool to achieve public goals, or with the "marketization" of public policy (Cerny 1992) that is driving the European varieties of capitalism closer to the American one (Hall and Soskice 2001). Public intervention then becomes the necessary trigger for private activity. For instance, in light of the complementarity between stock markets and venture capital (VC) funds, policy-makers may have promoted the emergence of stock markets for growing companies as a way to enlarge the exit opportunities for VC investments. This would be in line with previous evidence showing that, in setting the stage for an active private equity industry, the involvement of public bodies was essential, both in the US (Lerner 1996) and in Europe (see the evidence confirming the "seeding hypothesis" by public VC: Leleux and Surlemont 2003; Beuselinck and Manigart 2007).

The second issue concerns what lies behind the choice of a certain market organizational form? With respect to stock markets for growing companies, key architectural parameters concern the admission requirements, the level of information disclosure, and the allocation of regulatory responsibilities. Light regulatory settings can run counter to the monopoly positions of national stock exchanges on order flow, and can harm their liquidity if too many risky companies are allowed to go public. However, setting the entry and information requirements too high may discourage participation by TBSFs altogether, against the public goal of providing market-based financial support to new innovative ventures. How to solve this trade-off is an issue needing discussion.

A third issue concerns the macroeconomic consequences of stock market regulation. One of the lessons that we have learnt from the most recent recessions is that global financial stability crucially relies upon careful regulatory oversight of the financial system. Such a lesson has been translated by policy-makers into tighter requirements for financial disclosure on public companies and intermediaries (e.g. the Sarbanes–Oxley Act of 2002) and stricter capital requirements on banks (as with the Basel II Accords). However, stock market segments for growing companies create opportunities for regulatory arbitrage, therefore leaving the global financial system open to flows of unsuitable companies and cash shells. The entailed risks are even more likely to materialize if regulatory oversight is outsourced to financial intermediaries, possibly due to conflicts of interest. At the same time, in countries with a highly skewed firm-size distribution, such as Italy, facilitating the access of small firms to public equity is seen as a way out of the macroeconomic recession (see Cardia 2010).

A fourth and final issue of interest involves the balance between social costs and benefits of market-based financial support to TBSFs. Is the quotation of TBSFs worth the public and social costs associated with their large failure rates? Santarelli and Vivarelli (2002) asked a similar question about subsidies to new entrants. Their answer was that subsidizing entry might distort the learning processes that drive the dynamics of start-ups, and may corrode the market shares of the most efficient firms, while the positive effects are likely to vanish once the subsidy expires. All of this would result in a waste of resources. One wonders whether lowering the access requirements to financial markets would trigger similar effects. Furthermore, it has been observed that architectural choices often rely upon the imitation of the organizational features of foreign markets – as is the case with European stock exchanges "copying" the Nasdaq. Yet, it is increasingly suggested that the evolution of financial systems is constrained by path dependencies, involving inertia in household saving patterns, investment behaviors, corporate ownership structures, industrial relations, welfare systems, education, and the organization of research activities (Bianco *et al.* 1997; Bebchuk and Roe 1999; Hölzl 2006; Vitols 2005). If so, different countries have little to learn from each other's experiences, and imitative architectural choices can impose heavy social costs if the complementary institutions (in the sense of Aoki 2001) have not adapted.

For the above-mentioned reasons, it is worthwhile reviewing the main facts on the performance of high-tech stock markets, and of AIM in particular. Section 2 outlines the historical evolution of the second-tier markets for growing companies, while Section 3 illustrates some taxonomies proposed in the literature, with the aim of uncovering insights on the determinants of market architecture. The existing empirical evidence on the performance of AIM is reviewed in Section 3. Section 4 concludes.

2 A historical outline

Within the European context, the earliest attempts to set up second-tier markets for growing firms date back to the late 1970s. The pioneering markets for TBSFs were based on the so-called feeder principle: their function was to select the most profitable young companies and feed them upwards to the official list. The quotation of TBSFs was favoured by low admission requirements and low information standards. Posner (2004, p. 6) reports an exhaustive list of the stock markets based on the feeder principle. Those early experiences were unsuccessful: most investors perceived that feeder markets housed only poorly performing companies and preferred to wait for the best ones to be promoted to the main market (Posner 2004, 2009). Most feeder markets did not survive the 1987 stock market crash (Mallin and Ow-Yong 1998; Weber and Posner 2000; Ritter 2003).

In 1993, the European Union passed the Investment Services Directive (ISD): legislation aimed at integrating national investment services, including stock exchanges, by extending the principle of mutual recognition to service providers. By virtue of the ISD, an exchange regulated in one European Union country could operate in another via electronic networks and computer terminals. This enabled the creation of a pan-European stock exchange for young high-tech companies, which was promoted by the European Commission together with the European Venture Capital Association (Weber and Posner 2000; Posner 2004). The new market, the Easdaq, was inaugurated in 1996. Since it took the Nasdaq as an organizational model, it coupled low entry requirements with strong informational standards.[3]

National exchanges regarded the forthcoming creation of the Easdaq as a threat: the risk that financial activity might migrate to the new pan-European stock market led most national exchanges to set up their own segments for growing companies at the domestic level. The London Stock Exchange anticipated this move by creating the Alternative Investment Market (AIM) in June 1995. The Paris Bourse responded in 1996 by inaugurating the Nouveau Marché, and in 1997 the Deutsche Boerse established the Neuer Markt. Finally, trading on the Italian Nuovo Mercato began in June 1999. The so-called New Markets (NM) were designed according to the Nasdaq organizational model, with the exception of the AIM, which was closer to the old feeders. Admission and listing requirements on NMs have been summarized and analyzed by Clatworthy and Peel (1998), Bottazzi and Da Rin (2002), Goergen *et al.* (2004), Burghof and Hunger (2004) and Posner (2009), among others. The NMs set milder requirements than the main markets regarding capitalization, profitability, pre-IPO shareholder equity, IPO value, free float and track record, but tighter information disclosure rules, such as appointing sponsors to certify the company's compliance with the financial requirements and offer supervision and advice in the quotation process and in communications to the regulatory authorities; appointing market makers who match buyers and sellers; providing accounting information in line with international standards; and complying with lock-up rules constraining the disposal of shares by insiders.

The historical evolution of the main European NMs is represented in Table 5.1, reporting the number of member companies and the capitalization (in million US$) of the AIM, the Neuer Markt, the Nouveau Marché, the Nuovo Mercato and, for comparison, the Nasdaq, between 1995 and 2008. As can be easily grasped, none of the European markets even comes close to matching the size of the Nasdaq. The New Markets created by national exchanges in continental Europe experienced fast growth in the early years only. The Neuer Markt soon came to be the leading high-tech exchange in Europe, reaching a capitalization peak of more than US$113 billion in 2000; in the same year, the Nouveau Marché overtook the AIM in capitalization terms, with the younger Nuovo Mercato almost catching up with them. All of this urged the London Stock Exchange to make the AIM rules more rigorous and to set up the TechMARK segment in 1999, aiming to allow clearer identification of innovative and R&D-intensive companies within the official list. Since then, the AIM has sought to broaden its focus to SMEs in general, also outside high-tech sectors; moreover, prior admission to the LSE was an eligibility requirement for the TechMARK.

As of 2001, the burst of the so-called Internet bubble hit all of the markets quite hard. Several companies were forced to admit that they could not meet the earning forecasts declared in the introduction prospectuses, and the ensuing bankruptcies contributed to a general downward trend in stock prices and capitalization, resulting in numerous de-listings and rare new IPOs. Between 2000 and 2002, the drop in capitalization was dramatic: −91% in the Neuer Markt and

-68% in the French and Italian NMs, less so in the Nasdaq (−44%) and in the AIM (−25%). The Neuer Markt and Easdaq ceased operations in 2003 (Table 5.1). In January 2005, the Paris Bourse replaced the first market, the second market and the French NM with a single official list (Eurolist by Euronext). The Italian NM was replaced in September 2005 by MTAX; the relevant legislation shows that admission requirements are now very similar to those in the main market.

After 2001, the corporate and accounting scandals that marred the stock markets, rooted in conflicts of interest and insufficient regulatory controls, pushed regulators to reinforce both the capital requirements on banks (see the Basel II Accords) and the requirements for financial disclosure on public companies and intermediaries – as with the Sarbanes–Oxley Act of 2002 (SOX) in the United States. The restrictions on credit supply implied by the former may have stimulated the demand for public equity by TBSFs, while European markets seem to be better suited to satisfy such greater demand, in light of the opportunities for regulatory arbitrage between the US and Europe induced by the SOX. At the same time, the process of harmonization of the investment services across the European Union, which commenced with the ISD, was further developed by means of the Financial Services Action Plan (FSAP) of 1999, the Lamfalussy process, and their main offspring, namely the Markets in Financial Intermediaries Directive (MiFID). Issued in 2004 and implemented in the EU countries in November 2007, MiFID challenged the dominant positions of the national stock exchanges in two ways. First, it repealed the concentration rule (Art. 14(3) of the 1993 ISD), according to which retail orders handled by financial intermediaries had to be executed on a regulated market. Second, it allowed other trading platforms (the so-called Multilateral Trading Facilities, or MTF) to compete with regulated markets for order flows.[4]

Against the background of a changing financial landscape, the AIM managed to survive and experienced accelerated growth in both capitalization and the number of listed companies. NYSE-Euronext reacted to the enhanced post-MiFID competitive environment in 2005 by setting up Alternext, a MTF closely modeled on the AIM (Davies 2008; Degryse 2009), while First North was inaugurated by Nasdaq-OMX in Denmark in 2005. The AIM model was also "exported" as Tokyo AIM (2009) through a joint-venture between the LSE and the Tokyo Stock Exchange, and as AIM Italy (2009) following the acquisition of Borsa Italiana by the LSE. After the implementation of MiFID in the UK, the stocks listed on the AIM started to be traded on Plus, a British MTF focusing on small caps. The increasing influence of the AIM regulatory model is witnessed also by recent developments in US stock markets, such as the birth of OTCQX (Mendoza 2008). Outside Europe, the TSX Venture Exchange (formerly Canadian Venture Exchange, established in 1999) stands out as yet another long-lasting second-tier market for growing companies, competing with private venture capital for micro-cap firms at the pre-revenue stage (Carpentier *et al.* 2010).

3 Taxonomies of market organizational forms

In this section, we describe the organizational forms of stock markets for growing companies by means of three taxonomies proposed in the literature, and we give some insights as to the determinants of such organizational forms.

3.1 Listing requirements, information disclosure, and regulatory responsibilities

In a rather general way, one can view the architecture of a stock market as being made up of building blocks. Specifically, a stock market features a listing process, an information disclosure process, a trading system or price determination process, an oversight process, a sanction process, a graduation process, and the exchange's governance structure. Not all stock markets include all such components. For instance, many MTFs are mere trading systems (e.g., Chi-X Europe) as they trade in stocks listed in regulated exchanges. Graduation is the process through which a company delists from a second-tier market and is admitted to a first-tier market, and, as such, it is absent from first-tier markets. The organizational innovations brought about by feeders in the 1970s and 1980s and by the New Markets in the 1990s mostly concerned the listing and information disclosure processes. The turn of the century has witnessed changes in the governance of stock exchanges, which are increasingly undergoing demutualization,[5] whereas MiFID pushes the national stock exchanges to improve their trading systems along several dimensions, such as execution price, speed and likelihood of execution, and likelihood of settlement – a goal sometimes attained by integrating the competitors' trading technologies into their own system – as with LSE's acquisition of Project Turquoise, a MTF based in London.

A few attempts have been made at classifying the architecture of markets for growing companies, focusing on the way the above-mentioned building blocks are arranged. A first taxonomy, based on the trading system, distinguishes between order-driven and quote-driven markets. Such a classification is not specific to second-tier markets, and has been widely studied in the literature on market microstructure (O'Hara 1995; Spulber 1999). A second taxonomy, introduced by Posner (2005, 2009), maps exchanges along two dimensions: barriers to listing and informational standards. Finally, the listing, disclosure, and sanction processes can be organized according to alternative regulatory models, inspired by the rules-based approach and the principles-based approach respectively (Rousseau 2007).[6]

3.1.1 Order-driven vs. quote-driven markets

The explicit consideration of trading rules in the analysis of stock markets highlights two major trading systems or price determination processes, yielding two polar market settings: price-driven markets, in which prices are set by market makers, and order-driven markets, where orders are stored in a book and matched to generate prices. The Nasdaq, followed by the LSE, is historically the main instance of a price-driven market, while countries in continental Europe and in Japan tend to use order-driven systems. Nowadays almost all stock exchanges have adopted a hybrid system, combining an electronic book with market makers. The main advantage of a market making system is the high liquidity provided by market makers (for a synthesis, see Madhavan 2002). Market makers allow reduction of the temporal imbalances in order flow by maintaining a presence (Grossman and Miller 1988; Venkataraman and Waisburd 2007), and contribute to filling gaps from unbalanced order arrivals (Demsetz 1968; Nimalendran and Petrella 2003). The liquidity provided by market makers can also be viewed as a public good with positive externalities (Gromb and Vayanos 2002). A more recent and original argument comes from Mao and Pagano (2011), who consider the market maker as a risk manager who can play a crucial role during periods of high volatility. Such beneficial effects on liquidity can be helpful, especially for low capitalization companies. On the other hand, the main disadvantages of a market governed by prices are the high transaction costs that are induced by the existence of a bid-ask spread (Pagano and Röell 1990; Neal 1992). Market making can be viewed as favorable for large institutional investors because they can negotiate prices within the market makers' spread (Madhavan 1993), hence retail investors may be discouraged, limiting the extent of the market. Ultimately, the discrete power of the market makers can favor non-competitive behaviors, as was the case in the Nasdaq during the mid-1980s (Christie and Schulz 1994; Barclay 1997; Huang and Stoll 1996; Kandel and Marx 1997; Dutta and Madhavan 1997, among others). Nasdaq market makers adopted numerous illicit practices such as price collusion and spread of private information (Guennif and Revest 2005). Ultimately, whether TBSFs benefit from market making or not, it depends on the balance between risk-mitigation effects and market power exercised by the market makers.

3.1.2 Barriers to listing and informational standards

The taxonomy proposed by Posner (2005) ideally maps markets in a two-dimensional space defined by the strength of listing requirements and by the degree of information disclosure. Information disclosure concerns the accounting standards (national or international), the obligation to publish quarterly reports, the ad hoc disclosure of significant news, and the use of the English language in company reports. The barriers to listing include minimal initial requirements in terms of capitalization, assets, equity capital, trading history, and free float. Traditional stock exchanges are characterized by high listing requirements and tight disclosure, whereas markets for growing companies usually impose low listing requirements. However, while informational disclosure was rather low in the so-called feeders, it was higher in the Nasdaq, Easdaq and the markets in the Euro.NM circuit.

Posner's emphasis on admission requirements and information disclosure builds on economic insights that are widely shared in the literature. The shares of small, relatively untested companies are unlikely to be floated if the companies have to comply with the strict admission requirements of the official lists. It is well understood that small and young high-tech enterprises are informationally opaque because of their short track record and the novelty of their businesses. Tight information standards aim to overcome such opacity. A company's decision to submit to tight information disclosure rules may nevertheless be interpreted as a signal of high entrepreneurial quality, possibly triggering positive selection dynamics. Further, the information disclosure process works as an interface between a national stock market and foreign players (be them issuers, intermediaries, or investors), since the adoption of accounting standards and the use of a widely diffused language makes information intelligible worldwide. By attracting the quotation of foreign TBSFs, tight information rules enlarge the pool of potential market entrants, thereby increasing the likelihood that the market attracts companies with high business potential. Also, greater compliance with international standards improves coverage by international analysts, possibly yielding beneficial reputation and risk-mitigation effects. By the same token, low information disclosure can discourage international investors, but quite interestingly it can serve the interests of the national shareholders and stakeholders against inflows of foreign capital, and can shield the national stock exchanges from mutual competition, preserving their positions as national monopolists on order flow. Instituting markets with low information standards is in line with the European tradition of stakeholder-oriented corporate governance systems (Charreaux and Desbrières 2001; Balling 2004). Whether TBSFs benefit from stock markets managed in the interest of stakeholders is an empirical question, but notice that other stakeholders (e.g. large and established companies) may have more influence on the goals pursued by the stock market.

3.1.3 Rules-based vs. principles-based regulation

A third and final taxonomy builds upon the theoretical concepts of rules-based and principles-based regulation (Verheij *et al.* 1998, Burgemeestre *et al.* 2009). In a rules-based regulatory system, the content of regulation is made up of general, abstract, and universal rules, defined ex-ante, i.e. before adoption and implementation of the regulatory process. Neither exceptions, nor discretionary behaviors are allowed. In "producing" the rules, the legislator duly takes into account the principles that lay behind the policy goals. In turn, principles give rise to reasons in favor and reasons against. The legislator then assigns a weight to each reason, and each rule is the outcome of a weighted comparison between pro- and con-reasons. In a principles-based regulatory system, however, the weights to the pro- and con-reasons are assigned by the regulator or by an auditor case by case, and case-specific circumstances are taken into account when determining the weights. Participants in the regulative process, therefore, hold discretionary power. Rules-based regulation prescribes how business should be conducted, whereas principles-based regulation leaves companies free to decide how to align their conduct to the policy goals (Walsh 2008).

Rousseau (2007) applies the rules-based vs. principles-based taxonomy to stock markets, with reference to the admission, oversight, and disclosure processes. In markets regulated by a rules-based approach, the listing requirements are objectively defined, and the listed companies have to comply with formal rules, while in the principles-based regulatory approach, assessing the suitability of a company is outsourced to specialized financial intermediaries who hold discretionary power in performing their assessment (see Rousseau 2007). Such "trust intermediaries" are called nominated advisers on AIM, sponsors on Alternext, and certified advisers on First North. Nasdaq, Easdaq, the markets belonging to the Euro.NM circuit, and TSXV fall into the rules-based camp.

The case of AIM is an interesting illustration of the principles-based approach. AIM requires that every company seeking admission appoints a Nominated Adviser (Nomad). "A Nomad must be a legal entity with at least two recent years of corporate finance practice, having at least four 'qualified' executives and at least three relevant transactions for the same two years" (AIM Rules for Nominated Advisers, Article 2, February 2007). Nomads have to assess whether a company seeking admission is suitable for quotation, and later to provide assistance in order to ensure that the companies respect their continuing obligations. Nomads have to rigorously examine the applicant's business and must understand in detail the applicants' activities: business plan, managerial structure, financial and legal status, and so forth. AIM recommends the Nomads to visit the applicant's site of operation and to employ external experts to analyze the firm's business (Mendoza, 2008, p. 301). Such a suitability assessment replaces the formal listing requirements set by traditional stock markets. Nomads may also play an important role in corporate governance decisions, by persuading their clients to satisfy certain standards. As a matter of fact, the AIM rules do not mandate the adoption of specific corporate governance structures, and unlike LSE companies, AIM companies are not required to comply with the UK Corporate Governance Code.

All in all, Nomads act as gatekeepers, advisers and, ultimately, regulators of AIM-listed companies. Consistent with such decentralization/privatization of regulatory oversight responsibilities, the securities traded on AIM are not regulated by the UK Listing Authority (UKLA), and they do not need to be consistent with the EU directives implemented in the UK. Admission documents are not pre-vetted by the exchange or UKLA in most circumstances. The discretionary power of the Nomads is at least partly compensated by the investors' ability to prosecute Nomads if they are misled.

The main advantages and disadvantages of a principles-based regulation emerge quite vividly. With a faster admission process and customized oversight and disclosure, companies going public on AIM save on transaction costs, which are otherwise disproportionately borne by small companies. On top of this, the Nomads can perform certification and coaching roles for the quoted companies, much like venture capitalists. Yet, with decentralized, discretionary admission and monitoring, the viability of the market ultimately rests on the competence and integrity of the sponsors/nominated advisers. For one, applying principles requires greater case-specific knowledge than applying rules, as well as deeper understanding of the interdependencies between different, possibly conflicting principles. Outsourcing the regulatory responsibilities to financial intermediaries may provide high-powered incentives towards building such necessary competences, but knowledge tacitness and bounded rationality could seriously compromise the effectiveness of regulatory oversight, however tailored. Moral hazard may arise, too, since Nomads are hired and paid by the companies that they monitor, and can at the same time act as brokers and auditors for them (Gerakos *et al.* 2011). Social ties between market participants are likely to be rather strong, since the majority of the AIM investor's base turns out to be made of "sophisticated investors" – more precisely, institutional investors and investors specialized in the AIM companies (Mendoza 2008; Gerakos *et al.*, 2011). Institutional investors and the Nomads usually know each other well: the admission mechanism often involves private placements, where few institutional investors are invited to acquire shares in the applicant firm before the IPO.

In rules-based regulation, it is the policy-maker who needs competence and information to design general, abstract, and universal rules, and he/she needs them in advance, before actual implementation. Subsequent, unforeseen events may make those rules ineffective vis-à-vis the attainment of the underlying policy goals. Moreover, political economy considerations may cast doubts on the efficient collection and use of information by regulators and policy-makers – and consequently, on the adequacy of the rules. Therefore, uncertainty and government failure are the main threats of rules-based regulation.

3.2 Insights on the determinants of market organization

In line with Fligstein's (1996, 2001) approach to markets as social and political constructions, historical works in the fields of Sociology and Economics have underlined the connections between politics and the rise of financial markets (Carruthers 1999; Davis and Neal 2005). The former contribution insists on the continued interactions between the political elites, public finance and the stock exchanges established in the rise of financial markets in the UK during the eighteenth century. The latter work showed to what extent incentives for innovation depended on the sources of income of the traders, which in turn were determined by the institutional context and memberships rules.[7] While Carruthers' analysis is really close to Fligstein's (1996) proposal that markets should be viewed as political constructions (Preda, 2007), both historical works share the view that national regulation and powerful groups of actors play a role in the design of a financial market.

Consistently, the history of European stock markets for growing companies provides a peculiar viewpoint on how the organizational forms can be described by the adaptive responses and interactions of public and private actors to the threats and opportunities posed by ever changing macroeconomic and technological environments. In particular, one can identify two main tendencies: on the one hand, the role of supranational financial reforms in the emergence of stock markets for growing companies, and, on the other hand, the role of the national stock exchanges – as parent organizations of the newly instituted markets[8] – in shaping their market architectures.

Posner's (2005) account of the emergence of high-tech stock markets in Europe is enlightening. Second-tier markets for growing companies were established in the 1970s and 1980s under political pressure to help SMEs go public. The design of such markets, however, was left to national stock exchanges, and reveals misalignment between policy goals and the private interests of the exchanges. As a matter of fact, in the design of their second-tier markets, the national stock exchanges had to satisfy only one constraint for compliance with the policy goals: the entry barriers had to be low, in order to allow flotation of young small caps. Given that, the exchanges had considerable degrees of freedom in designing the other "building blocks". Such architectural choices were made in order to avoid two main risks. First, quoting small and untested companies might have scared investors, due to the implied information asymmetries, in turn harming the liquidity of large caps. Second, although the opacity of TBSFs might have been mitigated through tight information disclosure rules, tight information standards were incompatible with the positions of the exchanges as national monopolies on order flow. Markets such as the USM (UK) and the Second Marché (France) adopted lax information standards, and proactively used the graduation process (or feeder mechanism) to make sure that the best second-tier companies would transfer to the main market.

The birth of Easdaq and the ensuing emergence of New Markets can be traced back to public influence, too, through at least two channels. For one, the project of setting up a pan-European stock market would not be feasible without the ISD, since Easdaq exploited the principle of mutual recognition and the enhanced opportunities for electronic trading permitted by the ISD. This highlights the role of the European legislature as enabler of market emergence. But, as reported by Posner (2005), public actors did more. In particular, European Commission (EC) officials from DG13 (Information and Innovation) and DG23 (SMEs, Enterprise) actively promoted Easdaq. The interests of European Commission officials were aligned with those of key private actors, such as the EVCA. In Posner's interpretation, the EC officials pursued the goal of integrating the national financial markets, and at the same time were eager to find solutions to the structural unemployment and technology slow-down problems incurred by the European economies. EVCA sought wider exit opportunities for the investments made by their associates, and believed that new exit opportunities could be better provided by a broad, international, liquid stock market tailored to the needs of high-tech companies with little trading history but high growth potential. The EC promoted political discourse linking job creation to venture capital investments, used its resources and encouragement to convince those EVCA members who were against the Easdaq project, provided financial support to two preliminary studies to the non-profit organization EASD, and gave subsidies for Easdaq's early times of operation (Posner 2005). Private initiatives by the national stock exchanges for the supply of new markets only surfaced when the threat of a publicly-subsidized pan-European stock market materialized. The national stock exchanges would probably have been better off without high-tech segments housing risky ventures, but because the other national exchanges set up their own high-tech segments, it was in their own interest to respond. One may envisage a sort of strategic complementarity between the stock exchanges' decisions to supply new markets.

Interestingly, all of the so-called New Markets apart from AIM adopted a market design tailored to closely follow the Nasdaq approach (tight information standards, low entry barriers). The reason why Nasdaq was chosen as the reference model is still disputed. Nasdaq was publicized as a vital factor in the American success with new technologies, such as biotechnology (see Coriat *et al.* 2003). This shed light on potential dynamic efficiency gains, but explanations rooted in the quest for political consensus have been advanced too (see Posner 2005).

However, given the feeders' inability to perform, Nasdaq must have appeared as the only credible alternative design regardless of its purported macroeconomic effects, a view that is compatible with an adaptive learning perspective. In line with a knowledge-based approach to institutional change, one could think of Easdaq and the New Markets as enacting imitation strategies, with the Nasdaq market architecture as the target. Not by chance, Posner (2005) speaks of "Nasdaq copies". Following Rivkin (2001), the likelihood of success of an imitation strategy is tuned by the complexity of the targeted market architecture. The relevant definition of complexity, here, is based on that provided by Simon (1962): market architecture is complex if it consists of many components and those components are highly interdependent, so that novelties imply an extensive reorganization of the whole architecture. Complexity implies causal ambiguity, namely the difficulty of fully understanding the causal connection between architectural features and market performance (liquidity, efficiency, viability). Therefore, highly complex architectures can withstand successful imitation attempts even when information about their components is open to public scrutiny, or when several ingredients are correctly reproduced by imitators. If such intuition is correct, the historical record of collapsing New Markets suggests that the building blocks of high-tech stock markets are richly interdependent, and shows that much of the knowledge required to imitate market architecture is tacit. Nonetheless, interactions between economic agents are framed not only by formal rules, but also by informal constraints. The informal practices and routines that have materialized in a particular market can hardly be codified, let alone the social relationships that lay behind the networks of actors (see White 1981; Baker 1984; Granovetter 1985 on financial markets). The broad, country-specific institutional context where a new market is implemented matters, too, with all its complementarities, which, however, may be rather elusive. With hindsight, the perspective of complexity could have warned the EC and the national exchanges against setting up Nasdaq-like high-tech segments.

The recent trend of AIM imitators (see Alternext) and of AIM-initiated markets (as with AIM Italy and Tokyo AIM) can be best understood by means of Winter and Szulanski's (2001) concept of replication strategy, and Rivkin's (2001) analysis of the comparative difficulties faced by imitators and replicators at various levels of complexity. Although AIM possesses an informational advantage with respect to its imitators, in the form of a better assessment of the original "template", Rivkin's (2001) analysis indicates that with high complexity, the replication of market architecture by AIM can be nearly as difficult as its imitation by competitors. It is only at moderate levels of complexity that the replicator reaps the benefits of its superior information on the template.

A concrete illustration of the difficulties in imitating a financial market has been provided in the case of the French New Market by Revest (2010). The French Market Authorities did not give enough weight to the Paris Stock Exchange history and to the skills and experience of the intermediaries. For instance, market making was a completely new function in Paris and intermediaries were not able to perform it in an effective way.

4 The AIM: success factors and threats

Since its inception, the AIM market capitalization has evolved from £82.2 million in 1995 to £81,276.8 million in April 2011, and money raised by companies has increased from £96.5 million to £2,088.2 million during that same period (Table 5.2). The number of listed companies has increased from ten in 1995 to 1,165 in April 2011. The AIM has proved to be more resilient than most other markets for growing companies after the Internet bubble. Examining the AIM list, one finds that while most companies are based in the UK, some of them have significant assets outside the UK and can be viewed as international companies (Bauer and Boritz 2006). In addition, a growing number of foreign firms – and, notably, US firms – have entered the market in the last decade. The number of international companies quintupled between the end of 2003 (60) and August 2006 (304). Between 2004 and 2008, the number of new listings was higher on the AIM than on the LSE main market and on the Nasdaq (Gerakos *et al.* 2011; see Table 5.3). Quite interestingly, AIM was also able to attract companies from the Main Market, which outnumbered the companies graduating from AIM to the Main Market (see also Jenkinson and Ramadorai 2008; Gerakos *et al.* 2011).

These figures and facts all testify to the success of AIM if seen in terms of capitalization growth and increasing market participation. One goal of this section is to review the main theoretical reasons behind such "popularity". Yet, such a process of dimensional growth of the market can threaten the achievement of a satisfactory "quality" of the listed companies and the market's ability to support TBSFs. We explore this issue by reviewing the results of empirical studies about the performance of AIM companies.

4.1 Why is the AIM so popular?

How come a market catering to relatively untested small caps managed to survive across nearly two decades constellated with financial bubbles, to experience dimensional growth, and to even attract an increasing flow of international companies? To date, this question has not yet been subjected to empirical scrutiny, but a number of theoretical explanations can be formulated, including regulatory arbitrage, diversification of the list across sectors, and location advantages. Such explanations should by no means be viewed as strict substitutes, since the underlying mechanisms may be at work at the same time, and may even reinforce one another.

4.1.1 Regulatory arbitrage

The concept of regulatory arbitrage refers to the fact that companies can exploit differences among regulatory regimes in order to increase their profits. The migration of IPO activity from tightly regulated to lightly regulated markets, and the AIM's ability to attract companies previously listed on the LSE Main Market and on US exchanges, could be interpreted as outcomes of regulatory arbitrage. In this view, regulatory arbitrage was triggered by factors that increased the comparative advantage of AIM as a provider of liquidity and trading services. AIM's low fees, fast admission procedure, and customized oversight may have increased its attractiveness, whereas the attractiveness of other stock markets may have been diminished by regulatory changes, such as the Sarbanes-Oxley Act (SOX) in 2002.

To begin with, AIM offers low admission costs and listing fees to public equity for small firms. The total cost of an AIM initial listing appears to be lower than a Nasdaq initial listing. According to estimates, the cost of a listing is equal to US$3,426,300 for the AIM against US$4,472,000 for the Nasdaq, based on the premise of a company seeking to launch a $50 million IPO on both AIM and Nasdaq (Mendoza 2008, p. 307). The direct ongoing listing costs are also limited for the AIM ($147,300) compared to the Nasdaq ($3,515,500), in the case of a completed $50 million IPO. In addition to the low admission and listing fees, the AIM guarantees a quick introduction procedure. The standard procedure takes around three to six months to complete (Rousseau, 2007). It involves the production of an admission document (a prospectus) that is not appraised by the LSE authorities. Since 2003, companies can also use the fast-track route, available for issuers that are already listed on recognized foreign exchanges. In that case, the company does not have to produce an admission document. Further cost savings are related to the customized regulatory oversight guaranteed by the Nomads (see Section 3.1.3 for more details).

The effects of the cost advantages implied by the peculiar AIM model are reinforced by regulatory reforms affecting the competitiveness of other markets. The purported role of SOX in the erosion of the US financial markets competitiveness and the migration of IPO activity towards non-US financial markets is a hotly debated topic. Opponents of SOX blame it for increasing the costs of listing on US exchanges, and are favorable to a lighter approach to securities regulation (Bloomberg-Schumer Report 2007; Mendoza 2008). In their view, US companies after SOX feel the presence of a heavier regulatory burden and prefer to go public on lightly regulated markets, such as the AIM (Piotroski and Srinivasan 2008; Coates 2007; Leuz 2007; see also the Report of the Committee on Capital Market Regulation, or Paulson Report, 2006). Such higher regulatory costs are imposed without regard to firm size, hence SOX seems to particularly stimulate the migration of smaller firms (Mendoza 2008). Consequently, the post-SOX improved attractiveness of AIM is rooted in novel US regulation that does not discriminate between small and large firms (Jenkinson and Ramadorai 2008; Hoque 2011; Doidge *et al.* 2010).

The downside of regulatory arbitrage is that the US companies attracted by AIM may be companies unable to comply with strict information requirements, and their move to a lighter regulatory setting would signal their lower chances of survival. Indeed, supporters of SOX claim that its higher costs are more than offset by the long-term benefits related to positive selection effects. In particular, SOX may exert a dissuasive effect upon fraudulent behaviors (Duarte *et al.* 2013). According to the regulatory bonding hypothesis, companies that do submit to high regulatory standards signal higher quality to their investors (Coffee 1999, 2002; Stulz 1999; Reese and Weisbach 2002; Doidge *et al.* 2004; Zingales 2007).

4.1.2 Diversification of the list across sectors

Diversification of the list across sectors has presumably allowed AIM to avoid the collapse undergone by the European New Markets in 2001–2002. The composition of the AIM listing may have diluted an otherwise overly risky list based on TBSFs, thereby attracting companies, intermediaries, and investments. Even during the Internet bubble, the AIM list was more diversified than that of the other European NM, which mainly housed companies from a narrow set of R&D-intensive sectors. For instance, by the end of 1999, more than 80% of the Easdaq companies belonged to technology-based sectors, such as software, electronics, IT, biotech and medical equipment, telecommunications, and specialized equipment (Charlesworth 2000). On the Nuovo Mercato, telecommunications accounted for the largest emission share (over 40%); biotech and IT had relevant weights too (Petrella 2001). Conversely, high-technology firms on the AIM have never accounted for more than 25% of market turnover (AIM Statistics; see also Mallin and Ow-Yong 2012), and over time AIM has broadened its scope to include companies from traditional sectors, such as mining, oil, gas and real estate (Mendoza 2008).[9]

4.1.3 Location advantages

Location-specific advantages provide yet another set of explanatory factors. In particular, AIM seems to enjoy positive spillovers from the London financial center. London attracts a large pool of sophisticated institutional investors, endowed with experience and competencies, who build close links with both the companies and the Nomads. Previous studies on cross-listing have shown the importance of analyst coverage and experienced specialists, as well as "be with your peers" effects for young high-technology companies (Blass and Yafeh 2001; Pagano *et al.* 2002). The attractiveness of AIM is also related to UK venture capital being the main venture capital industry in Europe, representing about 21% of the European venture capital industry in 2009 (EVCA statistics; Revest and Sapio 2012). On top of this, it must also be noted that, over time, UK venture capitalists slightly improved their attitude towards young and small firms, while in the early 1990s they tended to give priority to large deal sizes and large companies (Murray 1999; Jeng and Wells 2000; Lockett *et al.* 2002; Mayer *et al.* 2005). A robust venture capital industry such as the one in the UK may have supported the growth of AIM: AIM offers exit opportunities to venture capital investments, and in return venture capitalists supply AIM with potentially fast growing companies that benefited from their advice.

4.2 The existing empirical evidence: Contrasting views on the performance of AIM

The increasing number of domestic and international firms listed, the continued capitalization growth, and the replication and imitation of the AIM model around the world all testify to the popularity of AIM. At the same time, one may envision a trade-off between the dimensional growth of the market and the average "quality" of the listed firms. Indeed, there are reasons to expect that the marginal entrant will be significantly worse than the extant firms in terms of long-term returns, growth rates, labour productivity, solvency, and survival rates. Conflicts of interest between the Nomads and the companies they supervise, rooted in AIM's discretionary screening and oversight approach, may attract companies that struggled to survive on the official list, as well as cash shells. Therefore, two questions naturally arise: is dimensional growth of the market achieved at the expense of listing quality? And what are the consequences for the market's ability to support TBSFs? Recent and ongoing empirical studies shed light on three main issues: the characteristics and performances of the firms listed on AIM, the reasons behind voluntary delistings, and the influence of the level of regulation on IPO migrations between AIM and the LSE Official List.

4.2.1 Characteristics and performances of the listed companies

A number of empirical works have examined the performance of the companies listed on AIM, with respect to operating performances, stock returns, survival rates, and corporate governance structures. The main concern behind those works is that facilitating the access to public equity can stimulate unsuitable companies to go public – and as those companies under-perform and leave the market, small investors suffer losses, and the public resources spent to support TBSFs get wasted.

Operating performances and stock returns. Hoque (2011) examines the operating performances of firms listed on the AIM and on the LSE Official

List. Over the period 1995–2010, the operating performance in the five years after IPO is, on average, negative for AIM for all years, and positive for the main market. Stock returns are lower on the AIM than on the Official List. AIM companies are less likely to make acquisitions, produce lower dividends, and are more likely to be cancelled. Yet, the probability of Seasoned Equity Offerings (SEOs) is higher on the AIM than on the Official List. Relatively bad operating performances by AIM firms are underlined by Gerakos *et al.* (2011), too, who perform a comparison between the AIM, the LSE Official List, the Nasdaq and the OTCBB. AIM post-IPO share prices are found to under-perform IPOs on other markets. AIM firms tend to under-perform even when they are accompanied by Nomads who are also brokers and auditors, and, more generally, the choice of Nomads does not have an influence on performance. Another test performed by the authors concerns the ability of the AIM to support the most promising firms. AIM's fast growing firms do not receive extra-returns: they are less likely to have extra positive outcomes than firms listing on other exchanges. Finally, the performance of AIM companies appears to be rather close to that of Pink Sheet companies as regards liquidity and informational asymmetry.

Survival rates. "I'm concerned that 30 per cent of issuers that list on AIM are gone in a year. That feels like a casino to me." This statement by Roel Campos, an SEC commissioner (Bawden and Waller 2007), has triggered two empirical studies asking whether the estimated survival rates of AIM companies are as low as suspected. Espenlaub *et al.* (2008) have found that AIM IPOs exhibit high survival rates over the short-term (one to two years after IPO), controlling for size, age, initial returns, country of incorporation, VC-backing, sector and time dummies. Over the long run (five years post IPO), the probability of surviving is positively related to age at IPO, size, and UK incorporation. Lastly, IPOs in information technology tend to have higher a probability of failure (over five years) than IPOs in other sectors. Consequently, the results by Espenlaub *et al.* contradict Campos's claim. Such optimistic results are contradicted by Gerakos *et al.* (2011), who show that AIM firms have higher instantaneous failures rates than firms listing on the Nasdaq, LSE Main Market, or the OTCBB: 24.8% of the AIM sample firms fail within a year of listing, quite strikingly in line with Campos's prediction. It must be said, however, that Gerakos *et al.* (2011) only control for market value at listing, sector fixed effects, and year fixed effects, therefore raising robustness issues.

Corporate governance. The topic of company quality is approached by Mallin and Ow-Yong (2012) through the analysis of the determinants of corporate governance voluntary disclosure on the AIM. Based on the examination of 300 financial reports, voluntary corporate governance disclosure is positively associated with company size, board size, turnover, and is higher for companies coming from the main market. At the same time, young companies with little or no trading history display a high commitment to corporate governance practices. Two main results can be emphasized: first, corporate governance practices largely depend on the internal characteristics of the AIM companies. Second, among the companies that are involved in good corporate governance practices, two different groups emerge. One group includes recently-established and young firms listed on the AIM which want to produce a positive signal toward investors. The other group contains mature firms that are used to advanced corporate governance practices or that have the required internal competencies.

4.2.2 Who are the switchers?

A comparison of the firms' listing changes between the AIM and the Official List of the LSE and their determinants has been performed by Jenkinson and Ramadorai (2008) and Campbell and Tabner (2011). Assessing the characteristics of the switching companies can be useful in order to understand whether attracting main market companies is beneficial or harmful for the average quality of the companies quoted on AIM. The quality of the companies switching from the Main Market to AIM can be inferred by exploring the reasons behind the switches or, alternatively, by estimating the effect of a listing change on the switchers' stock returns.

Concerning the determinants of the switches, the reasons quoted by the managers to explain their switch from the main market to the AIM appear to be consistent with the specificity of the AIM's model: lighter regulation, low listing costs, and a market design more favorable to small firms (Campbell and Tabner 2011). Among other reasons, it is worth mentioning the willingness to attract new shareholders. A particular firm profile emerges: companies that switch from the LSE Official List to AIM are rather small and credit constrained, do not generate enough cash flow, and suffer from too rigid and costly regulation. A shift to the AIM allows them to raise equity through Seasoned Equity Offerings, which are more numerous on the AIM, partly because firms on the AIM do not need shareholders approval for these transactions (Hoque 2011).

When the movements are from the AIM toward the main market, the predominant explanatory factors can be split in two interdependent categories (Campbell and Tabner 2011). First, companies may desire to increase their shareholders base. The empirical evidence shows that ownership concentration is declining when there is a switch from the AIM to the Main Market. In addition, listing on the main market may result in a more desirable profile and reputation, and hence in greater ability to attract new investors and talented personnel. Second, the market for corporate control seems more dynamic in the main market than in the AIM (Hoque 2011; Campbell and Tabner 2011). Consequently, companies may believe that they will have more opportunities to realize mergers and acquisitions in the main market. Campbell and Tabner (2011) also shape the agency-based hypothesis that managers choose to go to the main market at the end of a period of good performance in order to maximize their personal wealth.

Jenkinson and Ramadorai (2008) and Campbell and Tabner (2011) have examined whether the decision to change listing status between the LSE Official List and the AIM affects stock returns. Even if the two studies exhibit a few technical differences (Table 5.4), they converge to the same results: firms that shift from the Official List to the AIM exhibit negative abnormal returns at the announcement and positive abnormal returns after the implementation. The reverse movement is observed when firms move from the AIM to the Official List: they are characterized by positive abnormal returns at the announcement and negative abnormal returns afterwards. The authors' interpretation is that a switch from AIM to the Official List signals that the company is ready to face higher standards of disclosure and corporate governance, justifying a lower cost of capital – in line with the bonding hypothesis. The increase in the bonding costs is outweighed by a lower cost of capital, and the switchers' stock prices rise. After the implementation, expected returns fall because the risk premium for investors is lower, taking agency risk into account (Campbell and Tabner 2011). In the case of a switch from the Official List to the AIM, investors estimate that the agency risk will be higher in the AIM than in the Official List (because of a decrease in the bonding costs and a higher cost of capital), so the switchers' stock prices fall. Then, after implementation, the abnormal positive returns can be interpreted as a reward to shareholders for bearing increased levels of agency risks. Overall, agency risk plays a crucial role in explaining the differential returns of firms in two regulation regimes, in the spirit of Skaife *et al.* (2004). Jenkinson and Ramadorai (2008) particularly insist upon the role of institutional investors. They observe a small increase in aggregate institutional holdings in the six months following a switch to the AIM, from 23% to nearly 25%. This suggests that institutional investors are not abandoning companies once they switch to the segment with lighter regulation. The conclusion is that there exist particular investor clienteles for the two markets. Overall, switches from AIM to the Main Market may deprive AIM of dynamic companies, while companies that switch from the Main Market to AIM are of rather low quality.

4.2.3 Voluntarily delisted companies and cash shells

As we have previously mentioned, the dimensional growth of a stock market can come at the expense of the quality of the listing. The positive trend in market admissions can threaten the AIM functioning inasmuch as an increasing number of foreign entrants is made up of offshore vehicles and cash shells. If investors perceive the market as increasingly risky, even the most promising TBSFs may collect much less funds than they may have hoped.

Cash shells – through reverse takeover operations and under the guise of investment companies – have abused the market, as shown by the scandals involving Regal Petroleum and Chariot (*The Lawyer*, 27 November). In an attempt to fix this problem, in April 2006 AIM introduced a new rulebook for Nomads in order to strengthen their eligibility criteria, define their responsibilities more clearly, and make them more accountable for the companies they introduce (Mendoza 2008). A new rule was also implemented in 2005, according to which cash shells that raise less than £3 million on listing and do not complete a deal within a given time receive a six-month trading suspension. Due to this ruling, 38 companies were suspended from trading in 2006 (Bauer and Boritz 2006). In 2007 a new scandal altered the AIM functioning: the Langbar International fraud, perhaps the biggest fraud discovered on the AIM thus far. Prosecutors showed that this company had none of the assets it declared (£375 million).

Kashefi-Pour and Lasfer (2011) estimate the impact of a firm's capital structure upon the probability to voluntarily delist, using a sample of AIM-listed companies. The findings suggest that the firms that choose to delist had, at the time of IPO, a higher leverage, a higher share of intangible assets, less growth opportunities and lower capital expenditure than those choosing to stay listed. Such evidence is in line with the theoretical expectations about companies that go dark and companies that go private, which were included among voluntary delisting in the paper. Companies that choose to go private feel a necessity to change: they may be inefficiently managed, under-leveraged or under-valued by the market (for studies on other financial markets see, for instance, Lehn and Poulsen 1989 and Lichtenberg and Siegel 1990). On the other hand, going dark appears to be an answer to financial difficulties and distress. Companies that choose to go dark are usually small companies characterized by poor performance and low growth, whose difficulties are worsened by stock market regulation. According to Leuz and Wysocki (2008), the decision to go dark may also be motivated by agency problems and insiders interests. For instance, managers can avoid dismissal due to poor performance by going dark. Also, managers can choose to go dark in order to extract private benefits or protect themselves against regulatory actions.

5 Conclusion

Our main objectives in this paper were to explore the historical evolution of European stock markets for growing companies, to illustrate their organizational forms, and to evaluate their performances. We focused on the AIM because of its long-lasting dimensional growth, which led it to be recognized as a possible organizational model of financial markets able to bridge the equity gap of TBSFs.

What have we learnt from the history of the European financial markets dedicated to TBSFs? A first result refers to the variety of the market architectures which succeeded one another. There is no unique model for the stock market for growing companies, but several ones, according to different criteria such as listing requirements, information disclosure and regulatory responsibilities. Each organizational architecture displays both advantages and drawbacks for TBSFs. Second, adaptation to technological and macroeconomic innovations by private and public actors seems to be a major driver behind the observed variety of market organization forms. Public bodies at the national and supranational level (especially the EC) as well as the national stock exchanges have played a crucial role in shaping market architecture. Yet, the variety of the actors involved can induce conflicting views and the market design may result from severe negotiations, or appear as a compromise. Third, the emergence and development of financial markets dedicated to TBSFs have to be examined through the regulatory changes that could stimulate or, at the opposite, slacken a market's expansion. Regulatory changes produce unintended consequences which can be judged as positive or negative for a particular market, such as the beneficial effects of SOX on the AIM. Fourth, the history of financial markets dedicated to TBSFs sheds light on the issues of replicating and imitating a market. In the case of the "New Markets" created during the 1990s in continental Europe, "copying" Nasdaq turned out to be a failing strategy. The complexity of market architectures, compounding many highly interdependent elements in ways that are hard to codify, makes replication and imitation of a market very difficult tasks.

In this context, how can we explain the expansion of the AIM? The AIM does not arise from the European wave of Nasdaq copies. We have suggested that the AIM's popularity relies on regulatory arbitrage induced by some regulatory changes (e.g. SOX) that made the AIM relatively more attractive – at least for those companies that view tight regulation not as a signal of financial soundness, but, rather, as a heavy burden. AIM's ability to expand also relies on diversification of its listing across sectors, which dilutes the risk attached to companies in high-tech industries, and has allowed the AIM to better withstand the consequences of the Internet bubble crash. Finally, the AIM benefits also from locational advantages, i.e. from positive externalities coming from the London financial center.

Nevertheless, a detailed analysis of the empirical performances of the AIM – and especially relating to the characteristics of the firms listed – reveals contrasted views and shadow zones. On the one hand, several arguments are advanced to explain the AIM's attractiveness, from the viewpoints of both issuers and investors. First, there exists a specific class of investors for this market who are rewarded for bearing higher levels of agency risk – as compared to the main market (Jenkinson and Ramadorai 2008). Secondly, the AIM offers flexible and inexpensive trading services and regulation to firms that are small and credit constrained, and that are unable to go public on the main market (Campbell and Tabner 2011; Hoque 2011). On the other hand, the empirical evidence raises the issue of the quality of the firms. Operational performance and stock returns on the AIM are shown to be negative over the medium term (Hoque 2011; Gerakos *et al.* 2011). AIM's companies underperform with respect to companies listed on the LSE main market and on the Nasdaq (Gerakos *et al.* 2011). Nevertheless, these results would have to be confirmed in the future, but the very characteristics of the AIM's market model which were at the origin of its expansion – light rules, close relationships within a small set of actors, Nomads and institutional investors, reputation effects – can turn into in harmful features for the long-term stability of the market. Scandals and pricing anomalies have already happened; cash shells and reverse takeovers engender increasing regulatory concerns. Will a market characterized by a principles-based regulatory approach be able to control and sanction fraudulent behaviors? This debate is all the more necessary as the AIM model spreads internationally. Policy-oriented analyses aimed at averting the above-mentioned threats will benefit from novel empirical evidence concerning the profiles of AIM investors, the types and motives of corporate control operations involving AIM-listed companies, and the impacts of different regulatory approaches on the long-term development of financial markets for growing companies.

Appendix: European second-tier stock markets

Table 5.1 Historical evolution of second-tier stock markets in Europe, 1995–2008, compared to the NASDAQ: end-of-year listed companies and capitalization (in million US$)

Listed companies	*AIM*	*Nouveau M.*	*Neuer M.*	*Nuovo M.*	*NASDAQ*
1995	121	–	–	–	n.a.
1996	252	18	–	–	5,556
1997	308	38	17	–	5,487
1998	312	81	64	–	5,068
1999	347	111	201	6	4,829
2000	524	118	339	40	4,734
2001	629	164	327	45	4,063
2002	704	153	264	45	3,649
2003	754	137	–	43	3,294
2004	1,021	128	–	40	3,229
2005	1,399	–	–	38	3,164
2006	1,634	–	–	–	3,133
2007	1,694	–	–	–	n.a.
2008	1,550	–	–	–	n.a.
Capitalization ($ million)	*AIM*	*Nouveau M.*	*Neuer M.*	*Nuovo M.*	*NASDAQ*
1995	3,670.3	1,519,939.8	–	–	–
1996	8,809.8	1,511,824.4	956.7	–	–
1997	9,420.8	1,737,509.7	1,655.9	n.a.	–
1998	7,411.7	2,243,734.0	5,069.5	46,636.0	–
1999	21,740.9	15,261.1	74,571.8	6,996.9	5,205,620.4
2000	21,824.8	22,791.3	113,596.8	20,811.2	3,597,085.9
2001	16,731.8	13,604.3	29,942.1	11,120.0	2,739,674.7
2002	16,262.2	7,243.8	10,341.7	6,706.0	1,994,494.0
2003	32,162.3	10,267.4	–	10,425.3	2,844,192.6
2004	61,233.3	10,753.0	–	9,071.8	3,532,912.0
2005	98,816.3	–	–	8,615.1	3,603,984.9
2006	177,978.1	–	–	–	3,865,003.6
2007	196,917.1	–	–	–	4,013,650.3
2008	56,113.5	–	–	–	n.a.

Sources: AIM Market Statistics, World Federation of Exchanges. AIM data have been converted into US$ using the average December US$–GBP exchange rates (source: New York Federal Reserve website). "n.a." = not available.

Table 5.2 Historical evolution of the AIM, 1995–2011. Capitalization and money raised in £m. The number of companies takes into account the number at the beginning of each year

Year	*Number of companies*			*Market value (£m)*	*Number of admissions*			*Money raised (£m)*		
	UK	*International*	*Total*		*UK*	*International*	*Total*	*New*	*Further*	*Total*
06/1995	10	0	10	82.2						
1995	118	3	121	2,382.4	120	3	123	71.2	25.3	96.5
1996	235	17	252	5,298.5	131	14	145	521.3	302.3	823.6
1997	286	22	308	5,655.1	100	7	107	341.5	250.2	691.7
1998	291	21	312	4,437.9	68	7	75	267.5	317.7	585.2
1999	325	22	347	13,468.5	96	6	102	333.7	600.2	933.9
2000	493	31	524	14,935.2	265	12	277	1,754.1	1,338.2	3,092.4
2001	587	42	629	11,607.2	162	15	177	593.1	535.3	1,128.4
2002	654	50	704	10,252.3	147	13	160	490.1	485.8	975.8
2003	694	60	754	18,358.5	146	16	162	1,095.4	999.7	2,095.2
2004	905	116	1,021	31,753.4	294	61	355	2,775.9	1,880.2	4,656.1
2005	1,179	220	1,399	56,618.5	399	120	519	6,461.2	2,481.2	8,942.4
2006	1,330	304	1,634	90,666.4	338	124	462	9,943.8	5,734.3	15,678.1
2007	1,347	347	1,694	97,561.0	197	87	284	6,581.1	9,602.8	16,183.9
2008	1,233	317	1,550	37,731.9	87	27	114	1,107.8	3,214.5	4,322.3
2009	1,052	241	1,293	56,632.0	30	6	36	740.4	4,861.1	5,601.6
2010	965	229	1,194	79,419.3	76	26	102	1,219.4	5,738.1	6,957.6
2011 to Apr	941	224	1,165	81,276.8	18	4	22	87.1	2,001.2	2,088.2

Source: London Stock Exchange.

Table 5.3 Comparison of new listings for firms that listed and raised capital on the AIM, the LSE Main Market and the NASDAQ from June 1995 through December 2008 and on the OTC Bulletin Board from January 1999 to December 2008

Year	*AIM*	*LSE Main Market*	*NASDAQ*	*OTCBB*
1995	14	10	165	0
1996	66	38	358	0
1997	45	38	296	0
1998	17	31	240	0
1999	38	19	379	135
2000	120	51	269	36
2001	68	9	62	37
2002	53	17	72	47
2003	54	9	66	33
2004	208	20	170	46
2005	280	23	135	75
2006	279	41	142	77
2007	179	50	137	74
2008	40	11	22	53
Total	1,461	367	2,513	613

Source: Gerakos, Lang and Maffett (2011).

Table 5.4 Decision to change listing status: a comparison between Jenkinson and Ramadorai (2008) and Campbell and Tabner (2011)

Authors (year)	*Jenkinson and Ramadorai (2008)*	*Campbell and Tabner (2011)*
Markets	LSE, AIM	LSE, AIM
Data	210 firms down from Official List to AIM 53 firms up from AIM to Official List	245 firms down from Official List to AIM 86 UK firms up from AIM to Official List
Years	1997–2006	1995/1996–2008
Source	LSE, Datastream	LSE, Datastream
Methodology	Event study Uses weekly returns plus a Carhartt (1997) four factor model	Event study Uses daily returns plus a benchmark returns model that controls for industry residual returns and the possible interaction between market risk and change of listing status
Research question	How much do equity investors value regulation?	Does the regulatory environment influence the cost of capital?
Main results	Official List to AIM: negative abnormal returns in the 3 weeks surrounding the announcement and positive abnormal returns in the 6 month period following the switch	Official List to AIM: negative abnormal returns at announcement, between announcement and implementation; positive abnormal returns after implementation
	AIM to Official List: positive abnormal returns in the 3 weeks surrounding the announcement and negative abnormal returns in the 6 month period following the switch	AIM to Official List: positive abnormal returns at announcement, between announcement and implementation and abnormal returns after implementation
Interpretation	There exist particular investors clienteles for the two markets	Agency risk responsible for changes in firms returns over the announcement and implementation period

Acknowledgments

The authors would like to thank the participants of the Symposium "Innovation and Finance: a future-oriented Design of Financial Markets", Lisbon Civic Forum, Universität Hohenheim, 20–21 October 2010, Stuttgart. Support from the INET (Institute for New Economic Thinking) inaugural grant #220 "The Evolutionary Paths Toward the Financial Abyss and the Endogenous Spread of Financial Shocks into the Real Economy" is gratefully acknowledged.

Notes

1 Valérie Revest, TRIANGLE, Université Lyon 2 & CEPN affiliated researcher, Université Paris XIII (France).
2 Alessandro Sapio, DES, Università di Napoli Parthenope (Italy) and LEM, Sant'Anna School of Advanced Studies, Pisa (Italy).
3 A discussion of the listing requirements and information disclosure rules in stock markets for growing companies is provided in Section 3.
4 The implementation of MiFID raises debates on the relative merits of the fragmentation and consolidation of financial markets. While this debate goes beyond the scope of our analysis, such concerns are by no means new. Hautcoeur *et al.* (2010) and Hautcoeur and Riva (2012) illustrate the case of the Paris Stock Exchange, where reforms in 1893 gave formal recognition to the unregulated segment (the "Coulisse") and challenged the order flow monopoly of the strictly regulated official market (the "Parquet"). The monopoly of official brokers on all transactions in Parquet-listed securities was restored in 1898.
5 Demutualization is the process by which a co-operative or a customer-owned mutual organization changes legal form to a joint stock company.
6 Such taxonomies explore complementary aspects of market functioning, and are partly overlapping.
7 The analysis focused on the rules that governed memberships during the emergence of the LSE, the NYSE and the Paris Bourse, from 1792 to 1914.
8 For instance, LSE is the parent organization of the AIM.
9 Financial bubbles can arise even in traditional sectors, as with the real estate bubble of 2007. Yet, high-tech activities are characterized by much higher substantive uncertainty that challenges fundamental valuation even in the absence of bubbles.

References

Allen, J. (1992), *Starting a technology business*. London: Pitman.

Aoki, M. (2001), *Toward a comparative institutional analysis*. Cambridge, MA: MIT Press.

Baker, W. (1984), "The social structure of a national securities market", *American Journal of Sociology*, 89, 775–811.

Balling, M. (2004), "Objectives and theoretical foundations of the European Commission 1999 action plan concerning the framework for financial markets", *Journal of Banking Regulation*, 5, 256–286.

Barclay, M. (1997), "Bid-ask spreads and the avoidance of odd-eighth quotes on NASDAQ: an examination of exchange listings", *Journal of Financial Economics*, 45, 35–60.

Bauer, W. and Boritz, J. (2006), "Report on the UK's Alternative Investment Market – AIM", Working Draft, School of Accountancy, University of Waterloo.

Bawden, T. and Waller, M. (2007), "London v New York: top US regulator attacks AIM casino", *The London Times*, March 9.

Bebchuk, L.A. and Roe, M.J. (1999), "A theory of path dependence in corporate ownership and governance", *Stanford Law Review*, 52, 775–808.

Beuselinck, C. and Manigart, S. (2007), "Public venture capital across Europe: A 15 year perspective." In G. Gregoriou, M. Kooli, and R. Kraeussl (Eds), *Venture capital in Europe*. Oxford: Elsevier Finance.

Bianco, M., Gerali, A. and Massaro, R. (1997), "Financial systems across 'developed countries': Convergence or path dependence?" *Research in Economics*, 51(3), 303–331.

Blass, A. and Yafeh, Y. (2001), "Vagabond shoes longing to stray: Why foreign firms list in the United States", *Journal of Banking and Finance,* 25, 555–572.

Bloomberg, M. and Schumer, C. (2007), "Sustaining New-York's and the US's global Financial Services Leadership", January 22. www.nyc.gov/html/om/pdf/ny_report_final.pdf.

Bottazzi, L. and Da Rin, M. (2002), "Venture capital in Europe and the financing of innovative companies". In G. De Menil, H. Sinn, and R. Portes (Eds), *Economic policy*. Oxford: Blackwell.

Bottazzi, L. and Da Rin, M. (2005), "Financing entrepreneurial firms in Europe: Facts, issues, and research agenda", in V. Kanniainen and C. Keuschnigg (Eds), *Venture capital, entrepreneurship, and public policy*. Cambridge, MA: MIT Press.

Burgemeestre, B., Hulstijn, J. and Yao-Hua, T. (2009), "Rules-based versus principle-based regulatory compliance", Proceedings of Jurix, IOS Press, Amsterdam, 37–46.

Burghof, H.P. and Hunger, A. (2004), "The Neuer Markt: an (Overly) Risky Asset of Germany's Financial System" in G. Giudici and P. Roosenboom (Eds), *The Rise and Fall of Europe's New Stock Markets*, *Advances in Financial Economics*, 10: 295–329.

Campbel,l K. and Tabner, I. (2011), "Bonding, firm value and liquidity: an analysis of migrations between the AIM and the Official List of the London Stock Exchange", available at SSRN: http://ssrn.com/abstract=1608403

Cardia, L. (2010), Annual Report of Consob (Italian financial market regulator).

Carpentier, C., L'Her, J.F. and Suret, J.M. (2010). "Stock exchanges markets for new ventures", *Journal of Business Venturing*, 25, 403–422.

Carruthers, B.G. (1999), *City of capital: politics and markets in the English financial revolution*, Princeton University Press.

Cerny, P. (1992), *The changing architecture of politics. Structure, agency and the future of the state*, London: Sage.

Charlesworth, R. (2000), "EASDAQ key statistics", www.fisd.net.

Charreaux, G. and Desbrières, P. (2001), "Corporate governance: stakeholder value versus shareholder value", *Journal of Management and Governance*, 5(2), 107–128.

Christie, W. and Schultz, P. (1994). "Why do NASDAQ market makers avoid odd-eighth quotes?" *Journal of Finance*, 49, 1813–1840.

Clatworthy, M. and Peel, M. (1998), "The characteristics of new equity markets for SMEs". *Journal of Small Business and Enterprise Development*, 5(1), 79–92.

Coates, IV J.C. (2007), "The goals and promise of the Sarbanes-Oxley Act", *Journal of Economic Perspectives*, 21(1), 91–116.

Coffee, J.C. (1999), "The future as history: the prospects for global convergence in corporate governance and its implications", *Northwestern University Law Review,* 93, 641–708.

Coffee, J.C. (2002), "Race to the Top?: the impact of cross-listings and stock market competition on international corporate governance". Working Paper. New York: Columbia University.

Coriat, B., Orsi, F. and Weinstein, O. (2003), "Does biotech reflect a new science-based innovation regime?" *Industry and Innovation*, 10(3), 231–253.

Davis, L. and Neal, L. (2005), "The evolution of the rules and regulations of the first emerging markets: the London, New York and Paris Stock Exchanges, 1792–1914", *The Quarterly Review of Economics and Finance*. 45, 296–311.

Davies, R. (2008), "MiFID and a Changing Competitive Landscape", available at SSRN: http://ssrn.com/abstract=1117232.

Degryse, H. (2009), "Competition between financial markets in Europe: what can be expected from MiFID?" *Financial Market and Portfolio Management*, 23(1), 93–103.

Demsetz, H. (1968), "The cost of transacting", *Quartely Journal of Economics*, 82(1), 33–53.

Doidge, C., Karolyi, G. and Stulz, R. (2004), "Why are foreign firms listed in the US worth more?" *Journal of Financial Economics*, 71, 205–238.

Doidge, C., Karolyi, G.A. and Stulz, R.M. (2010), "Why do foreign firms leave US equity markets? *The Journal of Finance*, 65(4), 1507–1553.

Duarte, J., Kon, K., Young, L.A. and Siegel, S. (2013), "The impact of the Sarbanes-Oxley Act on shareholders and managers of foreign firms", *Review of Finance*, forthcoming.

Dutta, P. and Madhavan, A. (1997), "Competition and collusion in dealers markets", *The Journal of Finance*, LII(1), 245–276.

Espenlaub, S., Khurshed, A. and Mohamed, A. (2008). "Is AIM a casino? A study of the survival of new listings on the UK Alternative Investment Market", Mimeo, Manchester Business School.

Fligstein, N. (1996), "Markets as Politics. A political-structural approach to market institutions", *American Sociological Review*, 61(4), pp. 656–673.

Fligstein, N. (2001), *The architecture of markets*, Princeton NJ, Princeton University Press.

Gerakos, J., Lang, M. and Maffett, M. (2011), "Listing choices and self-regulation: the experience of the AIM", The University of Chicago Booth School of Business, unpublished paper, available at SSRN: http://ssrn.com/abstract=1739137

Goergen, M., Khurshed, A., McCahery, J. and Renneboog, L. (2004), "The rise and fall of the European new markets: On the short and long-term performance of high tech initial public offerings", In J. McCahery and L. Renneboog (Eds), *Venture capital contracting and the valuation of high tech firms*. Oxford: Oxford University Press.

Granovetter, M. (1985), "Economic action and social structure: the problem of embeddedness", *American Journal of Sociology*, 91(3), 481–510.

Gromb, D. and Vayanos, D. (2002), "Equilibrium and welfare in markets with financially constrained arbitrageurs", *Journal of Financial Economics*, 66, 361–407.

Grossman, S.J. and Miller, M.H. (1988), "Liquidity and market structure", *Journal of Finance*, 43, 617–633.

Guennif, S. and Revest, V. (2005), "Social structure and reputation: The NASDAQ case study", *Socio-Economic Review*, 3, 417–436.

Hall, P.A. and Soskice, D. (2001), *Varieties of capitalism: the institutional foundations of comparative advantage*, Oxford University Press.

Hautcoeur, P.C. and Riva, A. (2012), "The Paris financial market in the nineteen century: 65(4), 1326–1353.

Hautcoeur, P.C., Rezaee, A. and Riva, A. (2010), "How to regulate a financial market?

The impact of the 1892–1898 regulatory reforms on the Paris Bourse", HAL-PSE Working Papers 2010–01, Paris School of Economics.

Holtz-Eakin, D. (2000), "Public policy toward entrepreneurship", *Small Business Economics,* 15(4), 283–291.

Hölzl, W. (2006), "Convergence of financial systems: an evolutionary perspective", *Journal of Institutional Economics*, 2(1), 67–90.

Hoque, H. (2011), "One size does not fit all: evidence from IPOs, subsequent corporate actions and delisting in AIM and Main Market", School of Business and Economics, Swansea University, W.P. SBE 2011/1.

Huang, R. and Stoll, H. (1996), "Dealer versus auction markets: a paired comparison of execution costs on NASDAQ and the NYSE", *Journal of Financial Economics*, 41, 313–357.

Jeng, L. and Wells, P. (2000), "The determinants of venture capital funding: evidence across countries", *Journal of Corporate Finance*, 6, 241–289.

Jenkinson, T. and Ramadorai, T. (2008), "Does one size fit all? The consequences of switching markets with different regulatory standards". European Corporate Governance Institute, Finance Working Paper n. 212/2008.

Kandel, E. and Marx, L. (1997), "NASDAQ market structure and spreads patterns", *Journal of Financial Economics*, 45(1), 61–90.

Kashefi-Pour, E. and Lasfer, M. (2011), "The impact of leverage on the delisting decision of AIM companies", available at SSRN: http://ssrn.com/abstract=1787010

Leleux, B. and Surlemont, B. (2003), "Public versus private venture capital: Seeding or crowding out? A pan-European analysis", *Journal of Business Venturing*, 18, 81–104.

Lehn, K. and Poulsen, A. (1989), "Free cash flows and stockholders gains in going private transactions", *Journal of Finance*, 44, 771–787.

Lerner, J. (1996), "The government as venture capitalist: the long-run impact of the SBIR program", NBER Working Paper 5753.

Leuz, C. (2007), "Was the Sarbanes–Oxley Act of 2002 really this costly? A discussion of evidence from event returns and going private decisions", *Journal of Accounting and Economics,* 44, 146–165.

Leuz, C. and Wysocki, P. (2008), "Economic consequences of financial reporting disclosure regulation: a review and suggestions for future research", available at SSRN: http://ssrn.com/abstract=1105398

Lichtenberg, F. and Siegel, D. (1990), "The effect of leverage buyouts on productivity and related aspects of firm behaviour", *Journal of Financial Economics*, 27, 165–94.

Litvintsev, S. (2009), "At what do venture capitalist AIM?" Duke Law School, Unpublished Paper, available at SSRN: http://ssrn.com/abstract=1677633.

Lockett, A., Murray, G. and Wright, M. (2002), "Do UK venture capitalists *still* have a bias against investment in new technology firms", *Research Policy*, 31(6), 1009–1030.

Madhavan, A. (1993), "Consolidation, fragmentation and the disclosure of trading information", *The Review of Financial Studies*, 8(3), 579–603.

Madhavan, A. (2002). "Market microstructure: A practitioner's guide". *Financial Analysts Journal* 58(5), 28–42.

Mallin, C. and Ow-Yong, K. (1998), "Corporate governance in small companies: the alternative investment market", *Corporate Governance*, 6(4), 224–232.

Mallin, C. and Ow-Yong, K. (2012), "Factors influencing corporate governance disclosures: evidence from Alternative Investment Market (AIM) companies in the UK", *The European Journal of Finance*, 18(6), 515–533.

Mao, W. and Pagano, M. (2011), "Specialists as risk managers: the competition between intermediated and non-intermediated markets", *Journal of Banking & Finance*, 35(1), 51–66.

Mayer, C., Schoors, K. and Yafeh, Y. (2005), "Sources of funds and investment activities of venture capital funds: evidence from Germany, Israel, Japan and the United Kingdom", *Journal of Corporate Finance*, 11(3), 586–608.

Mendoza, J.M. (2008), "Securities regulation in low-tier listing venues: The rise of the Alternative Investment Market", *Fordham Journal of Corporate & Finance Law*, 13(2), 259–328.

Murray, G. (1999), "Early-stage venture capital funds, scale economies and public support", *Venture Capital*, 1(4), 351–384.

Neal, R. (1992), "A comparison of transaction costs between competitive market makers and specialist market structures", *Journal of Business*, 65(3), 317–334.

Nimalendran, M. and Petrella, G. (2003), "Do 'thinly-traded' stocks benefit from specialist intervention?" *Journal of Banking and Finance*, 27(9), 1823–54.

O'Hara, M. (1995), *Market Microstructure Theory*, Blackwell Business.

Pagano, M. and Röell, A. (1990), "Trading systems in European Stock Exchanges: current performance and policy options", *Economic Policy*, April, 63–115.

Pagano, M., Roell, A. and Zechner, J. (2002), "The geography of equity listings: Why do companies list abroad?" *Journal of Finance*, 57(6): 2651–2694.

Paulson Report – Report of the Committee on Capital Market Regulation (2006), www.capmktsreg.org/pdfs/11.30Committee_Interim_ReportREV2.pdf.

Petrella, G. (2001), "Sistemi finanziari e finanziamento delle imprese innovative: profili teorici evidenze empiriche dall'Europa", *Quaderni ref* n.4.

Piotroski, J. and Srinivasan, D. (2008), "Regulation and bonding: the Sarbanes-Oxley Act and the flow of international listings", *Journal of Accounting Research*, 46(2), 383–425.

Posner, E. (2004), "Copying the Nasdaq stock market in Europe: Supranational politics and the convergence-divergence debate", Working Paper PEIF-16.

Posner, E. (2005), "Stock exchange competition and the Nasdaq bargain in Europe", in: C. Parsons and N. Jabko (Eds), *With US or against US?: European trends in American perspective*, Oxford University Press.

Posner, E. (2009), *The origins of Europe's new stock markets*. Cambridge, MA: Harvard University Press.

Preda, A. (2007), "The sociological approach to financial markets", *Journal of Economic Surveys*, 21(3), 506–533.

Reese, W. and Weisbach, M. (2002), "Protection of Minority Shareholder Interests, Cross-Listings in the United States, and Subsequent Equity Offerings." *Journal of Financial Economics*, 66, 65–104.

Revest, V. (2010), "The failure of the French new market and the dynamics of rules", in M. Harvey, (Ed.), *Markets rules and the institutions of exchange*. Manchester: The Manchester University Press.

Revest, V. and Sapio, S. (2012), "Financing technology-based small firms in Europe: what do we know?" *Small Business Economics*, 39(1), 179–205

Ritter, J.R. (2003), "Differences between European and American IPO markets", *European Financial Management*, 9, 421–434.

Rivkin, J.W. (2001), "Reproducing knowledge: replication without imitation at moderate complexity", *Organization Science*, 12(3), 274–293.

Rousseau, S. (2007), "London calling?: The experience of the Alternative Investment Market and the competitiveness of Canadian stock exchanges", *Banking & Finance Law Review*, 23(1), 51–104.

Santarelli, E. and Vivarelli, M. (2002), "Is subsidizing entry an optimal policy?" *Industrial and Corporate Change*, 11(1), 39–52.

Simon, H. (1962), "The architecture of complexity", Proceedings of the American Philosophical Society, 106, 467–482.

Skaife, H., Collins, D. and Lafond, R. (2004), "Corporate governance and the cost of equity capital", available at SSRN: http://ssrn.com/abstract=639681

Spulber, D. (1999), *Market microstructure: intermediaries and the theory of the firm*, Cambridge University Press.

Stulz, R. (1999), "Globalization of Equity Markets and the Cost of Capital." *Journal of Applied Corporate Finance*, 12, 8–25.

Verheij, B., Hage, J. and Van Den Herik, V. (1998), "An integrated view on rules and principles", *Artificial Intelligence and Law*, 6, 3–26.

Venkataraman, K. and Waisburd, A. (2007), "The value of designated market maker", *Journal of Financial and Quantitative Analysis*, 42(3), 735–758.

Vitols, S. (2005), "Changes in Germany's bank-based financial system: A varieties of capitalism perspective", *Corporate Governance: An International Review*, 13(3), 386–396.

Zingales, L. (2007), "Is the US capital market losing its competitive edge?" Finance Working Paper, n°192/2007; available at SSRN: http:/ssrn.com/abstractid=1028701

Walsh, C. (2008), "Announcements and the role of policy guidance", *Federal Reserve Bank of St Louis Review,* 90(4), 421–42.

Weber, S. and Posner, E. (2000). "Creating a pan-European equity market: the origins of EASDAQ", *Review of International Political Economy*, 7(4), 529–573.

White, H. (1981), "Where do markets come from?" *American Journal of Sociology,* 87, November, 517–547.

Winter, S. and Szulanski, G. (2001), "Replication as strategy", *Organization Science*, 12(6), 730–743.

6 Financing instruments to combat the crisis

Ryszard Wilczynski

1 Global financial stability as a public good

Recent decades have witnessed a strong development of the financial sector, increasingly affecting the real economy. The surge in quantity, quality and diversity of financial market participants, instruments and procedures, as well as of resources, is indeed remarkable. Numerous studies have shown that, over time, financial development has been increasingly driven by two major factors: innovation and globalization. A wave of financial innovations has helped multiply the volume of available funding. Simultaneously, the process of globalizing the financial markets has progressed at an extraordinary space. It has been fueled by factors of different natures, such as the IT revolution, capital account liberalization, deregulation of financial sectors and cross-national expansion of huge financial institutions.

Financial development feeds into the macroeconomic performance of individual countries and of the global economy. There is broad agreement that financial stability is desirable for economic growth. In particular, international financial stability is perceived as a non-rival, non-excludable public good (Kanbur 2002). As such, it should be pursued by all stakeholders with an interest in maintaining it, as a lack of stability entails undesirable costs for them. The issue of which system is the most conducive to the achievement of international financial stability as well as for reaping benefits to economic growth from financial development is, however, more debatable.

One view being advocated is that international financial stability is, by and large, achieved via a system of competitive financial markets, which provides an adequate environment for independent policies of actors such as sovereigns and profit-maximizing investors. Such a framework helps avoid excess volatilities and positively affects growth. The concept of an efficient market can be traced back to the work of Samuelson (1965). A market-clearing mechanism appears to be instrumental to the efficiency of allocation.

Within that approach, financial innovations and financial globalization are regarded as important facilitators of economic growth. Under the market system, incentives for innovations, such as profit-maximizing or risk-optimizing behavior, are strong. New and improved financial products, procedures and institutions pave the way for the expansion of economic activity by facilitating the process of trading and risk-pooling, by raising the percentage of savings available for investment, and by improving the allocation of capital and producing higher total factor productivity (McKinnon 1973; Levine 1997). These positive effects are enhanced when financial innovations are allowed to work worldwide. Free financial movements are considered fundamentally sound. An improved diversification of risk and better allocation of capital are beneficial not only to growth, but also to the stability of consumption (Fischer 1998; Summers 2000; Henry 2007).

A contrasting view rests on the assumption of imperfect and inefficient markets. Inefficiency is attributed to such distorting factors as information asymmetries, herding behavior or principal-agent problems (Greenwald and Stiglitz 1986). Both financial innovations and financial globalization are producing and channeling considerable risks for the stability of the international financial system. In particular, the synchronized and simultaneous behavior of large fractions of market participants within the global financial network multiplies the effects of single decisions and leads to a systemic risk of excessive price volatilities.

New financial products are complex and not well understood by all market participants, thereby reducing transparency and contributing to the underpricing of risk (Dorrucci and McKay 2011). Securitization and sales of products to distant investors bring about a weakening of lending standards. Fast profit-seeking and innovative financial institutions attempt to lower their costs and, therefore, expand activities into the least regulated environment.

It is further recognized that financial globalization is aggravating risks by producing cross-border externalities and spreading them worldwide. These developments activate and fuel the process of financial contagion across the globe.

It is therefore argued that the system of free global financial markets tends to produce excessive volatilities when markets switch from euphoria to excessive pessimism which cannot be appropriately priced through a pattern of the risk premium and returns (Griffith-Jones 2003). In particular, international financial flows may not result in the equalizing of currency-adjusted and risk-adjusted rates of return everywhere. Abrupt, market-led adjustments undermine the stability of the international financial system.

In small open economies different causes of financial stress are usually translated into problems in the balance of payments. This refers to both external factors, such as terms of trade shocks or sudden reversals of capital flows, and domestic factors, which include excessive budget financing needs or run-on deposits. External shocks are more likely to lead to balance of payments tensions in economies with non-reserve currencies, even if sound policies are pursued. However, large fiscal deficits and the accumulation of domestic debt may also add to external debt which is further increased by the usually depreciating currency.

Recent decades have seen many balance of payments crisis episodes in emerging economies, from Latin America in the 1980s and 1990s, through East Asia, Russia and Brazil in the 1990s to Argentina, Turkey, Ecuador, Nigeria,

Venezuela and several European economies over the 2000s. Increasing reversals in international capital flows have posed persistent risk to external balances in emerging economies. Some of these crises have ended as sovereign defaults. In the aftermath of the global financial crisis, however, some advanced economies, particularly from the Eurozone periphery, were also affected by serious sovereign external problems.

Financial imbalances tend to expand, usually at high speed, across sectors and countries. In 2007 the crisis in the local financial subsector initiated developments that have diffused externally, creating, over time, fears of a global currency war.

The view challenging the efficiency of financial markets calls for an inevitable and active policy involvement required to address the balance of payment crisis. Financial contagion, which is speeding up and producing many considerable externalities, may create problems which cannot be dealt with exclusively by individual countries. Approaching international financial stability as a global public good implies, therefore, that coordinated, international actions are more effective than the market mechanism as instruments of response to a financial crisis.

2 Addressing financial crisis

Private financing is usually atomistic, uncoordinated and does not affect the debtor's policies which might otherwise have been desirable for crisis management. In a crisis environment, private creditors become increasingly reluctant to provide money and tend to rush for the exit.

In order to combat the crisis, the private sector may also undertake a number of actions other than direct financial contributions. Private financial institutions can improve their rules and procedures designed to reduce and prevent excessive volatilities on financial markets. The areas for improvement include risk management, disclosure and transparency procedures, valuation of assets, ratings of creditworthiness, and compensation policies, as well as shifts in governance (from sales-driven to more risk-focused). Market discipline is expected to enforce these improvements in an adequate and timely fashion. However, this process is not perfect.

In these circumstances, there is a case for a policy-driven response to a crisis. Such a response should be comprehensive, internationally coordinated and should encompass the main components, including macroeconomic policies, financial regulation and supervision, institutional frameworks and, finally, financing instruments.

Macroeconomic policies, particularly in systemically important countries, produce considerable externalities. If not sound, these policies may be a source of international financial contagion. It is, therefore, of critical significance that policies are adequate not only from the domestic perspective but also when being examined and assessed in the context of international spillovers. It is also important that macroeconomic policies include issues of financial stability, such as spillovers from financial markets to the real economy or combined risk across sectors.

Failures in regulation and supervision are among the widely accepted major causes of the global financial crisis. Some financial subsectors and many innovative instruments were not prudently supervised. Extending the scope of regulation and supervision, as well as the strengthening thereof, is of utmost importance. The process should include cross-border arrangements and be designed to solve home-host supervisory issues.

Financial crises give rise to adjustments within existing international financial institutions, establishing new ones as well as strengthening cooperation between existing ones. Reforms of the IMF policies, governance and resources have usually been accelerated in response to the crisis. The creation of the G20 and the Financial Stability Forum (transformed into the Financial Stability Board) were also consequences of crises. In response to the global financial crisis, the responsibilities of the G20 for international financial supervision and for the international monetary system have been broadened and deepened through establishing specialized working groups, while the coordination among the G20, the IMF, the FSB and the OECD on international financial stability was further developed.

For a successful response to a crisis, adequate financing instruments are also needed. With disappearing private financing, official financing or co-financing remains critical. Official instruments may, however, bring effects which go far beyond providing or catalyzing the liquidity needed for the correction of external balance sheets. These effects influence not only domestic balance sheets, but also all market participants, including through signaling. For example, external funds for a country in trouble may help improve its budgetary balance and encourage reforms. As far as external funds help to prevent a default for this country, other countries also benefit as they avoid an increase in global interest rates (Clark and Huang 2001). Adequate signals sent to market participants are instrumental in optimizing allocation and avoiding excessive volatilities of financial flows. All of these developments also help to strengthen the international monetary and financial systems.

Financing instruments can be classified into several groups. First, apart from direct official transfers to the debtors, there are market-based instruments which are sensitive to market prices. Second, financing instruments are designed for crisis resolution or – in precautionary form – for its prevention. Third, depending on the source of financing, instruments can be divided into national, regional and global. Finally, there are instruments addressing liquidity or solvency crises. A liquidity crisis is associated with the temporary problems of a country in raising funds. Problems are caused by external factors (e.g., by diminishing investor confidence) and may be solved via external financing. A solvency crisis means that the country, due to a deterioration in fundamentals or a severe protracted external shock, is unable to fulfill its external obligations unless new money and policy adjustment, including debt restructuring, is in place. Thanks to external financing, several cases of liquidity crises in recent years, such as those in Korea or Turkey, did not end with sovereign defaults, whereas in other cases, such as in Russia, Argentina or Uruguay, external support combined with policy adjustments did not prevent sovereign default. All in all, forty sovereign defaults have been registered over the period 1991–2004 (Borensztein and Panizza 2008).

However, the analytical distinction between liquidity and solvency crises has several shortcomings. Extreme cases include countries with strong fundamentals and adequate policies but which are nevertheless facing problems with access to financing (liquidity crisis) on the one hand and sovereign default (solvency crisis) on the other. Between the two extremes there are intermediate cases, as protracted and increasing liquidity problems may turn into a solvency crisis. Furthermore, the perception of solvency depends on the ability and the will of policymakers to implement future adjustments (Ortiz 2002).

It is also reasonable that financing instruments conform to certain requirements. Generating financial flows when access to market financing is problematic is obviously the central issue. In that context, it is useful for instruments to be innovative, i.e., include new, non-traditional ways of raising public funds or complementing them by leveraging private financing or enhancing existing public financing frameworks. However, the amount of financing has to be optimized, not necessarily maximized. It should respect the country's external financing needs, e.g., the size of its short-term debt, as well as the sustainability of the financial position of its creditors. The amount and conditions of financing should also help avoid moral hazard problems on the side of both debtors and creditors.

Significant volatility and sudden reversals on financial markets call for the flexibility of financing instruments. In particular, it is important that in emergency cases high and prompt access to official financing is possible. Flexibility of instruments should go hand-in-hand with their predictability.

In order to be feasible, financing instruments have to respect fair burden-sharing on the side of creditors as well as fair distributional effects on the side of debtors. It is also significant that institutions providing international financing are perceived as credible ones, otherwise efficient financing instruments may remain unused.

3 Selected financing instruments: their review and assessment

Demand from countries for official external financing has significantly fluctuated over time and the toolkit of financing instruments has been sensitive to these fluctuations. Episodes of financial instability usually gave rise to new financing facilities and to the enhancement of existing ones, thus expanding the set of standard instruments such as bilateral or multilateral loans (e.g., stand-by loans from the IMF have recently been enhanced). These episodes of financial turmoil were preceded by sustained periods of easy credit, ample international flows, overlending to many countries and the building-up of financial imbalances.

The original response to the sovereign external crisis of the 1980s was the Baker Plan (1985), under which new official financing as well as loans from commercial banks to developing countries in trouble combined with structural reforms would allow those countries to spark economic growth and to grow out of their debts. This strategy lacked predictability and innovativeness, and it failed as it was designed to address liquidity crises, whereas the problems experienced by a number of developing countries turned out to be solvency-related issues.

Unlike the Baker Plan, the proposals that followed included debt forgiveness. The Robinson Plan (1988) proposed global securitization of debt of developing countries through a new agency – the International Debt Facility. Under the Mitterrand Plan (1988), and through SDR allocation, the new international institution would guarantee future interest payments to those creditor banks which accept debt reduction. The proposals included new financing schemes, but were not implemented.

The most comprehensive strategy designed to address the solvency crises of developing countries came in 1989 as the Brady Plan. Creditor banks could exchange non-performing loans for new Brady bonds issued by debtor governments, with debt relief but also with debt guarantees. Debt relief operations were collateralized with US Treasury zero-coupon bonds to be financed from new conditional loans from international financial institutions to developing countries or from their reserves. As a result, creditor banks received new lower, but performing and guaranteed, assets, while debtor countries arrived at reduced external debt.

The Brady Plan included a mix of official and market financing instruments and was innovative in several respects. The official claims of creditor banks were replaced with Brady bonds which were tradable on the secondary market, allowing banks to remove old debts from their balance sheets. Coordination among creditors and their acceptance of some losses paved the way for new lending. Lenders were offered an array of options permitting the operation of debt restructuring to be tailored to their various preferences. Those options included two types of Brady bonds: par bonds and discount bonds. Additionally, creditors could opt for a limited amount of non-discounted, debt conversion bonds combined with the commitment to provide new lending (new money bonds) to debtor countries. The fourth option was a cash buyback – i.e., a repurchase of debt at a specific price. The buybacks were to be financed by international financial institutions.

In 1994 Poland restructured its external commercial debt of US$14 billion under the Brady Plan, with a 50% debt relief; 25% of the debt principal was bought-back with a discount from creditors. Of the remaining part, creditors opted for 30-year discount bonds (61% of the total) and par bonds (10%), as well as for new money bonds (5%). Brady bonds were collateralized with zero-coupon US Treasury bonds (bond principal) and with cash (bond interest), both set aside on the escrow account. In combination with the restructuring of the official debt which took place, on comparable terms, in 1991, the Brady arrangement turned out to be instrumental to Poland's return to a manageable external balance and to regaining access to financial markets.

By the mid-1990s 16 developing countries, predominantly in Latin America, less frequently in Eastern Europe, Asia, Middle East and Africa, had restructured over US$200 billion of their total commercial debt under the Brady Plan, with the amount of debt forgiven totaling US$65 billion (Arslanalp and Henry 2007).

The overall assessments of the Brady Plan are, however, less clear-cut. In particular, the financial industry has argued that official financing within the Plan was not sufficient, and that the predictability of debt forgiveness encouraged developing countries to accumulate debt to commercial banks (Vasquez 1996).

The Brady Plan has, to some extent, overshadowed debt conversions – standard market instruments also designed to alleviate the problem of servicing external sovereign debt. These instruments assumed that the debt would move from hard to local currencies which were subsequently used for local facilities, either by the original creditor or by a third party that bought the original claims at a discount on the secondary market. Depending on the sector of investment, different types of debt swaps were introduced, with debt-to-equity and debt-for-nature and development swaps dominating the scene.

After the introduction of the Paris Club debt swap clause in 1990, debt conversions also gained the contribution of official financing to debt management operations. Creditor governments were allowed to increase the amount of debt forgiveness provided that debtor governments transferred local currency to a financial facility for investment in the agreed sector. As these debt conversions were initiated collectively under the Paris Club, ultimately they were implemented on a voluntary, bilateral basis. In Poland the EcoFund has been established to select projects for nature conservation and environmental protection and to finance these projects with resources from debt-for-nature swaps agreed with several Paris Club creditors.

In the aftermath of large crisis episodes (i.e., in the early 1990s, 2000s and 2010s), international financial flows have rebounded and, over time, gained momentum towards big waves. In these circumstances, the policy challenge was to manage the surge in flows in order to avoid their possible reversals, financial crises and contagion. In particular, emerging economies were attracting foreign capital as they were increasingly becoming an important part of global resilience rather than an independent source of international volatility.

The theory on the policy intervention into capital flows is not conclusive and reflects the diverging views on the (in)efficiency of financial markets. Some empirical studies covering periods of the 1990s (Williamson *et al.* 2003) and 2000s (Ostry *et al.* 2010) suggest that such interventions could have helped strengthen the resilience of economies towards crises.

In order to manage surges in capital flows, tax-type instruments have been implemented in a number of emerging economies, predominantly in Latin America and Asia. A mandatory unremunerated reserve requirement on capital inflows which was effective in Chile in the 1990s, as well as in Columbia and Thailand in the 1990s and 2000s, is among widely explored tools. Conceptually, the reserve requirement was not innovative as it could be traced back to 1972 and a Tobin-tax proposal for a levy on international currency transactions. In

Chile the reserve was required for one year, regardless of the duration of the capital inflow, so short-term flows were more heavily taxed than long-term ones. It is usually recognized that these tax-type instruments have helped alter the maturity of inflows, rather than reduce their magnitude, as well as alleviate pressures on monetary and exchange rate policies (Ostry *et al.* 2010).

Expecting a surge in portfolio inflows after the global financial crisis, several emerging economies have implemented different measures for flow management: e.g., a withholding tax on non-residents' purchases of bonds (Korea, Thailand), tax on non-resident portfolio equity and debt inflows (Brazil), an increase in reserve requirements on foreign currency deposits (Indonesia) or loans (Peru), and a capital gains tax on non-residents' investments in the stock market (Peru). As the effects of these instruments are still to be fully explored, it may be argued that they should be temporary and second-tier tools, after adequate macroeconomic and macroprudential policies.

A wave of emerging market crises of the late 1990s/early 2000s gave a push to the development of national and regional financing arrangements because global instruments, such as IMF facilities, had at that time been regarded with some criticism and disappointment by countries in crisis. It was not accidental that the two new and, to some extent, innovative IMF instruments designed for crisis prevention (Contingency Credit Line – CCL) and resolution (Sovereign Debt Restructuring Mechanism – SDRM) were either not implemented or have expired unused.

Instead, the decade of the 2000s saw an unprecedented build up of official reserve holdings, particularly in emerging economies. Self-insurance against the risk of a balance of payments crisis was the driver (albeit not exclusively) of the process. Global foreign exchange reserves grew from c.a. US$1.8 trillion in 2000 through US$4 trillion in 2005 and up to US$8.1 trillion at the end of 2009, i.e., to 14% of the total global GDP. Dramatic increases of reserves occurred, particularly in emerging Asia, and included, but were not limited to, economies hit by financial crises, while regional reserves have exceeded an estimated precautionary range. Asian central banks account for a bulk of world reserve holdings and eight of them are among the ten largest reserve holders.

Official reserve holdings provide a standard cushion against adversity for an economy exposed to external shocks. They allow for prompt, flexible, predictable and unconditional access to resources in case of need. However, these reserves not only entail costs which cannot be ignored, but are often found to be excessive (Mateos y Lago *et al.* 2009).

Foreign official assets fuel Sovereign Wealth Funds (SWF) – government investment vehicles managed separately from reserves. Assets under the management of SWF are estimated at US$3–4 trillion and are basically allocated with a view to maximizing long term returns. This, however, does not exclude investments for balance of payments purposes, so SWF may complement reserves in their task to mitigate external shocks.

In parallel with the accumulation of official reserves, over the 2000s regional financing arrangements have developed, mostly in Europe and Asia. In the European Union the new balance of payment assistance facility (Medium Term

Financial Assistance – MTFA) replaced the earlier instrument in 2002. Under MTFA the amount of €12.5 billion was set aside to assist non-euro EU member states facing balance of payments difficulties. The assistance would take the form of conditional loans with terms decided by the Council, with the co-financing of and in coordination with the IMF.

In 2000, thirteen Asian countries launched the Chiang Mai Initiative (CMI), an arrangement designed to address balance of payments difficulties in the region. The total amount of US$78 billion was envisaged to be disbursed through a system of bilateral swap arrangements under which local currencies were swapped with US dollars. In the case of each disbursement, the first 20% of the amount was unconditional, whereas an IMF program was required for the remaining 80%.

Equipped with such a system of facilities and resources designed to work towards international financial stability, in 2007 the global economy entered financial turmoil, contagion and an overall crisis. The severity of the crisis gave rise to the development of a financial toolkit as existing instruments and funds turned out to be insufficient. First, a balance of payments crisis hit a number of emerging economies, mostly in Europe. However, later on, an international contagion gained momentum.

A state of emergency made it barely possible to work out and implement new instruments. Therefore, first responses to the global crisis came down mostly to expanding financing potential under the existing set of instruments. From November 2008 to March 2009 the IMF signed 14 stand-by arrangements. Driven by G20, the Fund has agreed on a tripling of its financial resources, including an additional SDR allocation equivalent to US$250 billion and an increase of over US$500 billion for the resources of the emergency backstop program New Arrangements to Borrow (NAB).

The EU has multiplied, first to €25 billion and finally to €50 billion, the MTFA resources. Hungary, Latvia and Romania have been approved for MTFA assistance under joint EU/IMF programs. Several EU governments have also contributed bilaterally. On another front, ten EU countries were working on levies on financial institutions designed to stabilize financial sectors in case of crisis risks. The process of harmonizing national bank resolution funds is under way at the EU level. In Asia the CMI has also increased its resources to US$120 billion.

Several central banks have set up and enhanced bilateral swap arrangements with their partners. From December 2007 the FED set up currency swap lines with other central banks of both advanced and emerging economies. Scandinavian central banks entered into swap arrangements with Iceland, whereas in Asia the central banks of China, Japan and Korea have set up swap lines with Indonesia, Malaysia and Hong Kong.

In 2009 the IMF launched a new financing instrument – the Flexible Credit Line (FCL). In fact, it was the global crisis which pushed Fund members to finalize a long-lasting discussion on a facility which was originally set up in 1999 as the CCL. With improved design and conditions, the FCL has found demand from Mexico, Columbia and Poland totaling US$78 billion under original arrangements. Designed primarily for crisis prevention, with ex ante conditionality, uncapped funds which are available upfront and with unrestricted renewals, the FCL is a flexible tool for which only strongest performers are eligible. In late 2010, the IMF approved the Precautionary Credit Line (PCL) – the FCL-type instrument for economies that do not qualify for the FCL.

Towards the end of 2009 emergency measures helped to bring some relief from the stress upon international financial markets. Soon thereafter, however, new weak spots in the international financial system emerged, as some economies from the Eurozone periphery have faced severe and increasing problems in their fiscal and external balance. A concerted rescue operation by the EU, in coordination with the IMF, included several new facilities:

- a pool of bilateral loans to Greece from Eurozone countries totaling €80 billion, with burden-sharing determined by their contributions to the paid-up capital of the ECB. The IMF provided €30 billion in the form of a stand-by program;
- European Financial Stabilization Mechanism (EFSM), collected by issuing instruments with a total value of up to €60 billion and with the EU budget guarantee, further on-lent to EU member states which experience, or are threatened with, severe financial disturbances caused by factors beyond their control;
- European Financial Stability Facility (EFSF), a temporary facility issuing its own debt instruments of up to €440 billion in total value, backed with the guarantee of the Euro-area member states and on-lent to member states facing financial difficulties.

In late 2010, both EFSM and EFSF served to financially support Ireland with the amount of €40 billion under the emergency loan program, complemented by financing from the IMF and by bilateral loans from the UK, Sweden and Denmark.

Despite the considerable effort which has already been undertaken to reduce international financial stress, policymakers should remain alert to current risks as well as continue to work on measures to stabilize the system in the medium- and long-term.

4 In perspective: towards multilateral and preventive instruments

Prior to the global financial crisis, national and regional instruments for crisis prevention and resolution formed a considerable part of the toolkit, giving rise to concerns over the malaise of global governance (Ahearne *et al.* 2006). However, the global crisis revealed a more pressing need for policymakers to go beyond independent, local initiatives and towards broad, international, collective and coherent actions.

National instruments, like official reserves, are subject to diminishing returns and entail considerable costs for a country in the form of opportunity, transaction or sterilization costs as well as at the global level in terms of inefficiencies of economies of scale and a contribution to large global imbalances and the associated risks. Globalizing reserves ought to reduce the total amount of reserves needed and, therefore, reduce the costs of accumulating and maintaining them. In a crisis-deflated global environment individual countries may be tempted to raise trade protection and undertake competitive devaluations, i.e. to take actions which are internationally destructive. A move to multilateral financing should have helped countries avoid such actions.

The advantages of globalizing reserves gain significance when the amount of reserve holdings is found to be excessive, i.e. reserves are being accumulated for other than precautionary reasons. Conventional benchmarks for reserve adequacy – the value of three months' of imports; 100% of short-term debt; 5–20% of broad money – are considerably exceeded in a number of emerging economies. According to these measures, precautionary motives for reserve accumulation account for only a part (about two-thirds) of total reserve holdings (Mateos y Lago *et al.* 2009).

Conventional benchmarks of reserve adequacy are, however, challenged as being too narrowly based and are being updated to reflect risk levels of different potential sources of balance of payments pressure and tailored to country-specific conditions. Within this updated approach, some emerging economies still turn out to have accumulated excessive reserves (IMF 2011).

Excessive reserves entail pricing distortions leading to inefficiencies at the international level. These inefficiencies are associated with risks of exchange rate misalignments, asset price bubbles, increased search for yield under ample global liquidity as well as with imbalanced trade flows. Official reserve holdings are heavily concentrated in terms of currency composition as well as geographically, with the largest reserve holders being among those countries which do not necessarily need them the most for self-insurance against the crisis. For these reasons, self-insurance may be found to be globally inefficient, and the need to rise multilateral financing instruments is aggravated. Nevertheless, the preference of countries, particularly in time of crisis, for national reserve holdings as the more prompt and automatic tool of funding, cannot be excluded.

Multilateral considerations have to be taken into account in case of instruments designed for capital controls which may generate various international spillovers. As capital controls help reduce volatile and risky capital flows, they may contribute positively to international financial stability. Simultaneously, it is argued that such measures introduce market distortions, protect the financial sector from external competition and, therefore, bring deteriorating effects for the allocation of capital across countries (Ostry *et al.* 2010). If capital controls crowd-out better policies, in systemically important economies in particular, externalities will be channeled to other countries. If currency is undervalued, instruments of control may allow policymakers to avoid the appreciation otherwise needed to address global imbalances. Such instruments introduced by one country might also encourage other countries to follow suit, particularly if capital flows are redirected to those countries. It is therefore desirable that capital controls are dealt with (e.g., monitored, assessed) collectively, with international financial institutions serving as useful platforms.

Considerable externalities and financial contagion, systemic risks as well as a public good approach to the international financial stability, represent a rationale for increased multilateral financing in order to combat the crisis. The complexity of the issues to be solved as well as the costs involved and the resources needed make private or bilateral funding hardly sufficient. Without multilateral funding the world's ability to provide stability for international finance would be limited.

In a parallel move, financing provided proactively by means of precautionary insurance-type instruments will gain significance. In the medium- to long-term more countries will expectedly strengthen their policies and gain better market access and, therefore, will need external support designed, to a large degree, to maintain market confidence and to cope with potential rather than actual risks.

The development of a toolkit aimed towards increasing multilateral and precautionary financing is under way. Besides the augmentation of the FCL and the introduction of the PCL, it is also proposed to multilateralize the FCL and to make it available for a group of countries hit by shocks at the same time. Providing funds to a group of countries would effectively address the first mover problem. There is also a proposal for the Multilateral Swap Line (MSL), as the bilateral lines extended thus far by large central banks have been temporary and selective, i.e., dependent upon trade and financial links between the delivering and recipient country. MSL would provide automatic access to financing to a group of countries if the risk were global and regional. Both proposals have raised a number of concerns from the IMF member countries. Alternatively, cross-border liquidity provision in the form of bilateral swap lines could be harmonized – under the auspices of the BIS – into cross-border collateral arrangements (CBCA's).

In Asia, CMI was multilateralized in early 2010 when the network of bilateral swaps was transformed into a single, uniform facility. Arguably, it has helped prevent regional volatility as no funds have been drawn.

In the EU, in 2013 the EFSF temporary rescue fund will be transformed into the European Stabilization Mechanism (ESM), a permanent facility designed to serve as the lender of last resort for sovereigns. In the face of the severity of the sovereign debt crisis in the Eurozone which, arguably, shows signs of a solvency rather than of a liquidity crisis, rescue operations in the form of loans should be complemented with the debt reduction mechanism.

The process leading towards more collective and preventive instruments has to cope with several challenges:

First, burden-sharing in multilateral financing. Under collective arrangements national contributions clash with the common benefit – international financial stability – which, as a global public good, is generally available to everyone. National considerations encourage countries to renounce agreed commitments and let the others bear the cost, i.e., to free-ride. Under the IMF arrangement of

Multinational Consultations on global imbalances, the action plan agreed upon by five countries turned out to be less effective than expected as, to a certain degree, national interests outweighed collective goals. The 2010 Greek bailout by the EMU also provides an example of negotiations with a brinkmanship game in the context of facing national costs and benefits against those at the European level (Fahrholz and Wójcik 2010). Each country participating in multilateral financing has to be persuaded and convinced to contribute its fair share of the total.

Second, a moral hazard problem associated with official insurance-type financing. This may lead to a suboptimal outcome on the part of private creditors who are tempted to reduce their diligence, as well as that of sovereign debtors who get an opportunity to dilute policy and institutional reform and incur larger debts. Insurance provides a cushion in case of a crisis and makes such behavior by borrowers and lenders more likely. Counterarguments to the moral hazard case include the reduction, through insurance, of the likelihood of tail risks. It is also argued that insurance may actually encourage reforms, as a financing cushion makes success in overcoming crisis more likely (Jeanne *et al.* 2008).

Mitigating risks related to moral hazard depends on the conditionality of financing instruments, which affects the eligibility of countries for these facilities as well as the predictability and automaticity of the access to funds. It is desirable for these instruments to include risk-adjusted pricing. Conditionality should help send market participants adequate signals on macroeconomic fundamentals and policies of sovereigns willing to use instruments. The FCL and the PCL, new IMF precautionary facilities, are perceived as sound screening and well-informing tools, whereas the signaling outcome of their predecessor (the CCL) has been subject to criticism.

Finally, the credibility of the institution providing funds. The IMF already provides multilateral financing in special circumstances where private creditors are reluctant to do so. The Fund would have to enhance the insurance role of its instruments. In order to effectively fulfill this role, the IMF should continue reforming its governance, including the system of voting and representation. The misalignment of the real weight in the global economy and finance and of the voting power among the membership is illustrated below (see Figure 6.1).

This misalignment will be further reduced, although not eliminated, by the reform quota agreed by the Fund members in 2010 but not yet effective. In parallel to the IMF, the G20 (including 19 countries plus the EU) is also initiating initiatives in multilateral financing. This influential group is designed to reflect the world's largest economies; however, arguably this is not adequately achieved – Spain, Taiwan, the Netherlands and Poland are outside the group, but could pretend to be in. In the medium- to long-term perspective, new developing economies with the largest growth potential – for example, those labeled as "Next 11" – will likely have to be accommodated into the circles of international decision makers. The adequacy of the international systems distributing power and representation would be reviewed and improved.

To sum up, in a crisis environment where private financing is limited, official funding assumes a significant role in order to alleviate financial stress as well as

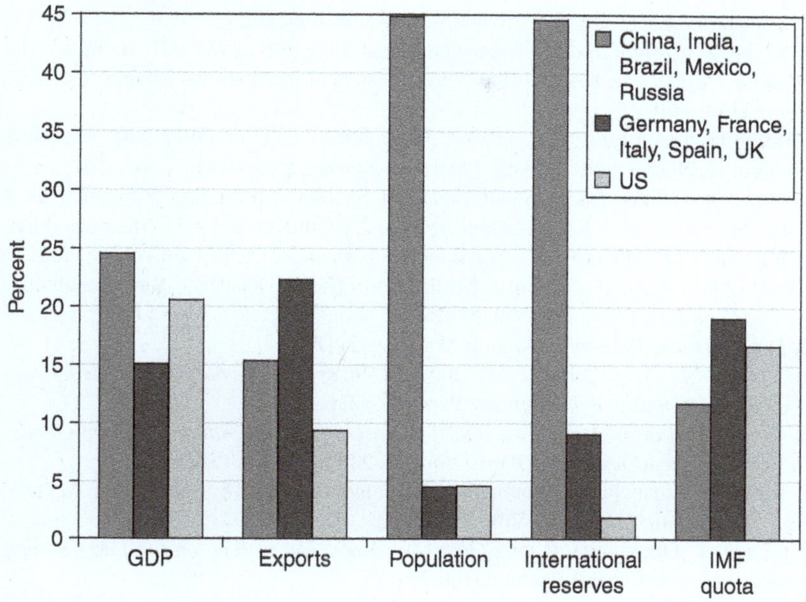

Figure 6.1 Weight of 11 countries in the global economy and finance and their representation at the IMF (2009; % of world/IMF total). (source: "Rebalancing Growth. World Economic Outlook April 2010" IMF, Washington DC, 2010).

initiate and catalyze adjustment. This role remains critical when official funds are designed to facilitate and enhance crisis prevention. From a global perspective, collective financing instruments exhibit advantages over national tools, provided that all of the involved participants remain committed to collective goals and the institution providing the funds is perceived as a credible source. Adequate conditionality attached to proactive and preventive financing helps to mitigate the risks associated with moral hazard.

References

Ahearne, A., Pisani-Ferry, J., Sapir, A. and Veron, N. (2006), *Global Governance. An Agenda for Europe*, Breugel Policy Brief No. 7.

Arslanalp, S. and Henry, P.B. (2003), *Is Debt Relief Efficient?*, Stanford Graduate School of Business Paper No. 1837.

Borensztein, E, and Panizza, U. (2008), *The Costs of Sovereign Debt*, IMF Working Paper No. 238.

Clark, P. and Huang, H. (2001), *International Financial Contagion and the IMF. A Theoretical Framework*, IMF Working Paper No. 137.

Dorucci, E. and McKay, J. (2011), *The Financial Crisis and the Strengthening of Global Policy Coordination*, ECB Monthly Bulletin, January.

Fahrholz, C. and Wójcik, C. (2010), *The Bail-Out. Positive Political Economics of Greek-type Crises in the EMU*, CASE Network Studies and Analyses, No. 413.

Fischer, S. (1998), *Capital Account Liberalization and the Role of the IMF* in: *Should the IMF Pursue Capital Account Convertibility?*, Essays in International Finance No. 207, Princeton University.

Greenwald, B.C. and Stiglitz, J.E. (1986), Externalities in Economies with Imperfect Information and Incomplete Markets, *Quarterly Journal of Economics*, Vol. 101.

Griffith-Jones, S. (2003), International Financial System and Market Efficiency as a Global Public Good, in: I. Kaul, P. Conceição, K. Le Goulven and R.U. Mendoza (eds), *Providing Global Public Goods – Managing Globalization*, Oxford University Press.

Henry, P.B. (2007), Capital Account Liberalization: Theory, Evidence and Speculation, *Journal of Economic Literature*, Vol. 45, No. 4.

IMF (2011), *Assessing Reserve Adequacy*, Washington DC..

Jeanne, O., Ostry, J. and Zettelmeyer, J. (2008), *A Theory of International Crisis lending and IMF Conditionality*, IMF Working Paper No. 236.

Kanbur, R. (2002), *International Financial Institutions and International Public Goods: Operational Implications for the World Bank*, G-24 Discussion Papers No. 19.

Levine, R. 1997, Financial development and Economic Growth: Views and Agenda, *Journal of Economic Literature*, Vol. 35. No. 2.

Mateos y Lago, I., Duttagupta, R. and Goyul, R. 2009, *The Debate on the International Monetary System*, IMF Staff Position Note No. 26.

McKinnon, R.I. (1973), *Money and Capital in Economic Development*, Brookings Institution, Washington DC.

Ortiz, G. (2002), *Recent Emerging Market Crises – What Have We Learned?*, Per Jacobson Lecture, July 7, Basel.

Ostry, J., Ghosh, A.R., Habermaier, K., Channou, M., Qureshi, M. and Reinhardt, D (2010), *Capital Inflows. The Role of Controls*, IMF Staff Position Note No. 4.

Summers, L.H. (2000), *International Financial Crises: Causes, Prevention and Cures*, American Economic Review Vol. 90, No. 2.

Samuelson, P.A. (1965), *Proof That Properly Anticipated Prices Fluctuate Randomly*, Industrial Management Review Vol. 6.

Vasquez, I. (1996), *The Brady Plan and Market-Based Solutions to Debt Crises*, Cato Journal Vol. 16, No. 2.

Williamson, J., Griffith-Jones, S. and Gottschalk, R. (2003), *Should Capital Controls Have a Place in the Future International Monetary System?*, A paper for a meeting of the International Monetary Convention held by Reinventing Bretton Woods Committee 13–14 May, Madrid.

7 Automotive dynamics in the Stockholm and southern German regional economies[1] – a comparison

Gunnar Eliasson

1 The new balance between small scale and volume[2] production

Industrial dynamics is an experimental economic phenomenon. Players of all sizes, controlling resources of various magnitudes and origin and being guided by more or less visionary ideas of what to do, clash in markets increasingly characterized by innovative product competition. Business mistakes abound, and the winners that disrupt the market are not necessarily the best ones. The current restructuring of global automotive markets, therefore, is a display of industrial dynamics that takes you close to the fundamental unpredictability of an *Experimentally Organized Economy* (EOE; Eliasson 2009, 2011).

I compare the sophisticated automotive competence bloc in Southern Germany (Bavaria/ Baden-Württemberg: B/B-W) with automotive production on the Swedish West coast, and in the Lake Mälar region around Stockholm. The focus will be on passenger cars. Since the same firms also manufacture trucks, buses and earth-moving equipment, and even aircraft, and since some experts have believed in strong synergies throughout all these production activities, I will have to address the entire transport vehicle industry.

Automotive markets have been in the midst of an experimental restructuring process for years, moved by new technology and global competition, with not only unpredictable, but also regionally frustrating outcomes. My focus will be on the Swedish automotive industry, and the risk that it may vanish altogether (like the Swedish shipyard industry, or, for that matter, the UK automotive industry) but also on the role, in its possible disappearance, of larger and more forceful German automotive manufacturers.

The diverse opinions on what is going on, and what needs to be done, among decision makers in the industry illustrate the uncertainty facing participating players. This is even more so for industrial economic analysts such as myself. So, I will leave it for the readers to judge for themselves what my documentation leads up to, rather than trying to point my analysis, through appropriate prior assumptions, in a particular predictive direction. My story will therefore also highlight the risk of using overly simplified single-valued strategic business models, based on traditional static economic theory in order to navigate in an economic/industrial world best characterized by immense, multidimensional and complex dynamics, notably:

a Volume manufacturing strategies to minimize costs and to compete with prices in world markets where customers increasingly demand product quality innovation;
b Suddenly abandoning your old competence base to rush into the New Markets of the New Economy where entirely new competence demands rule profit performance.

There will also be an opportunity to illustrate the empirical relevance of some of the assumptions figuring in so-called new growth theory.

My discussion is carried out against the backdrop of the radical change in engineering occasioned by the merge of computing and communications (C&C) technologies, developed in a parallel paper (Eliasson 2011), in which it was observed how a new balance between small scale and volume production is being established and an entrepreneurial-based renaissance of engineering was taking place. There are four levels of inquiry:

1 New product design technology increasingly offers increasing returns to product development (R&D) that reduce the need to raise volumes to cover escalating product development costs, as is suggested by the "new growth model" (of Jones and Williams 1998), and
2 The distributed manufacturing technology that drives globalization of production, captures new systems productivities, as was suggested in Romer's (1986) "new" macro growth model, or for that matter, already elaborated on in the micro macro economics of Marshall's (1890, 1919) industrial districts[3].
3 When looked at from a micro perspective, *networking externalities are achieved when distributing manufacturing over markets for specialized subcontractors*, as observed in Eliasson (2003). Such distributed production is more flexible, and also opens up direct access to the low wages of the not-yet industrialized world.
4 Business management systems, that may make complex hierarchies more transparent, but also more inflexible, encourage a volume mentality.

I argue that when *single-valued strategic decision models* (based on static economic theory and outdated management experience from the past) *address multidimensional dynamic business problems*, the experimental nature of dynamic markets come into full flower, raising the business failure rate. The winds of change are so forceful that even entire industrial regions are at peril.

The gradually emerging distributed solutions to production organization are, however, also offering enormous productivity gains that will fundamentally and permanently change the global production structure. We may talk about a renaissance of mechanical engineering.[4]

I find that what holds back the immediate exploitation of the industrial productivity potential of new C&C technology, and the transition to a new globally distributed production organization in a particular region or economy are (1) a lack of requisite competences on the part of business management; (2) the raised risks of management failure in a now more complex and unfamiliar business opportunities space; (3) an institutional environment in the industrial economies that discourages entrepreneurs to act on the opportunities; and, not least, (4) a general political aversion among the (still) rich industrial economies to absorb the unpredictable reshuffling of monetary wealth, individual welfare and political power that unavoidably accompanies a successful transition of that kind. The economic incentives driving change are, however, so large that the experimental transition to a new global organization of production will continue. The automotive industry itself plays an intriguing role in the reorganization of engineering, its products being the "vehicle" for change, and the infrastructure of roads a containing bottle-neck in the full exploitation of distributed production. (It is easy to understand how the limited capacity of European roads to accommodate such flows may prevent the full realization of positive networking externalities associated with distributed production. All automobile travelers on the congested German autobahns have experienced how bottlenecks, uneven traffic rhythms and unpredictable arrival times not only reduce positive externalities, but also contribute to energy waste and pollution.)

Taken together, the extreme complexity governing the experimental business dynamics makes the long-term outcome in terms of success and failure analytically indeterminate. The history of the emergence of the global automotive industry, its post-WWII fragmentation, its current enormous size and its political leverage where it is currently located, and the constantly ongoing technological transformation make a perfect example. As production distributed over markets of specialized subcontractors delivers a much larger production value with a significantly smaller input of workers, the local employment of unskilled workers becomes a growing concern. Since a number of these production units were earlier internal parts of a large firm that have now been separated and distributed over markets as autonomous firms/subcontractors that can access the entire global market, each (of those which survive) can now benefit from significantly larger economies of scale. A radically *different balance between small scale and volume production* is gradually being established.

With firms and entire regions at peril the industrial transformation becomes a policy problem and worries have been expressed about the future existence of the automotive industry in Sweden. A consequent question therefore is: Are there realistic policies that will prevent the disappearance of the Swedish automotive industry? Or is the optimal policy strategy to opt out and do something else? On this I have already argued in a companion paper (Eliasson 2011) that the engineering industry is far from an old mature industry ready to be phased out of the rich industrial world, but is, rather, going through a renaissance. So, it may even be that the reason for business failure in Sweden is not signaling the end of a hopelessly old industry, but rather management minds locked into strategic business models that worked better in the past (Eliasson 2005b).

2 The experimental dynamics of new industry formation

The transport equipment industry, and the automobile industry in particular, has been rapidly adopting the new C&C technologies in three principally different ways: C&C technology has (1) revolutionized the products, (2) fundamentally changed the ways firm hierarchies are organized and managed (Eliasson 2005b), and (3) revolutionized the design, development and manufacturing processes by making the integration of globally distributed value chains over markets of specialized subcontractors both technically reliable and commercially viable.

The convergence on a new (automotive) industrial organization is, however, messy and governed at the ownership levels over imperfect asset markets by actors that have not only shareholder value on their minds, but also by management visions that are not necessarily loaded by great industrial insights. *As single-valued strategic visions clash in multidimensional global markets, what we call an experimental process of industrial change is generated*, the distributional outcomes of which are close to being unpredictable.

The role of the demanding customer

In the long run no more advanced products than there are customers who understand how to use them, and their value to them, and who are able and willing to pay will be developed. Competence bloc theory makes a point of connecting innovation and commercializing agents with the customer (Eliasson 2009: 2; 2010: 43). This is particularly problematic for automotive industries, where most of the value chain of technical innovation, commercialization and manufacturing has long been integrated within one giant company. Breaking it up and distributing it over market of autonomous and specialized subcontractors, still remaining in top down control is a new management experience that requires new innovative competences.

The more varied customer demands, and the more varied the product characteristics that producers are capable of developing, the more difficult it becomes for competing suppliers to figure out both what will be demanded, and what competitors will do on the basis of similar expectations, and be sufficiently flexible to adjust supplies to what they expect customers to want just then. Business risks increase to the extent that a single product mistake may wreck the entire company. Producers therefore have to organize their product development as business experiments, and be prepared often to be very wrong. The automotive industry carries all these characteristics. While the upper end of its product range exhibits the attributes of a fashion industry that integrates transport efficiency and comfort with status, luxury and pleasure, the low end tends to market usefulness in large volumes at minimum prices.

Articulated strategies may be necessary at the top of large hierarchies, but if too naive and imposed top-down these may not only stifle innovation, but also orient the entire business in the wrong direction. *Business strategies*, or visions *are always compositions of simplified notions of what to do*, based on hunches of what the customer wants; expectations about competitors' expectations, and structured to convince skeptical and not very knowledgeable financiers and conservative boards, and also to please the visionary himself. To convincingly present an innovative product for unprepared outsiders, or conservative top management in a large firm, is difficult. It can be theoretically demonstrated that the absence of innovation should be typical of large hierarchies. Their top-down strategic implementation substitutes for hands-on customer closeness. Dearden *et al.* (1990) refer to the Soviet Union as an extreme case of the failure to adopt innovations even from a rich supply of potentially innovative technologies. Okamuro *et al.* (2011), in comparing the administered economy of Japan and the more entrepreneurial economy of the Netherlands, find statistical support for the stifling impact on entrepreneurship of centrally imposed policies. The negative effects of lost opportunities on long-term macro growth were also captured in simulation experiments on the Swedish micro to macro model (Antonov and Trofinov 1993).

To capture the preferences of fickle upper end customers, intuition is normally superior to industrial logic. The iPhone could not have been conceived by Nokia, or been passed through its management filter (*Ny Teknik* 31 March 2010, Nr 13–14: 13ff.). Similarly, Ericsson was lucky to have a mobile telephone skunk works financed over a secret military account, hidden and protected from its top management, which tried its best in the early 1980s to kill it (Eliasson 2010: 99ff.). Saab Automobile converted its 900 Turbo model to a cab version in 1986, without top management knowing about it before it became public. The cab became a commercial success, contrary to Saab´s standard product line.

Circumstances like the above allow articulate politicians to climb to the top of large public corporations, with large resources. Business men and engineers who run the small operations, with small or no financial reserves to cover up failure, do not talk as much. It is, however, fortunate if business strategies vary considerably since some may be rather right. The positive consequences for the entire industry of not losing the winners outweigh the incidence on total industry cost structure in the form of the business mistakes (a transactions cost) that are typical of an Experimentally Organized Economy (EOE, Eliasson and Eliasson 2005a). Those costs are ruled out by assumption in the mainstream economic model, which therefore is a faulty tool to use in industrial analysis for both business analysts and policy advisers. Business mistakes should in fact be regarded as a *normal cost of economic development* that have to be accepted to minimize the loss of winner (Eliasson 2009). By eliminating business mistakes through effective controls of visible costs, the unexpected business successes that make such a difference for long-run growth also tend to be eliminated (Eliasson 2005a; Antonov and Trofinov 1993). Since statistically visible costs are easier to control for in business information and management systems than the invisible costs firms incur in the form of lost winners, modern Enterprise Resource Planning (ERP) systems not only make the large businesses that use them more conservative. They also foster a volume and cost preoccupation with management that makes them vulnerable, and may take their business in the wrong direction (see further companion paper, Eliasson 2011). I will keep this possibility in mind when we get more concrete about the automotive industry.

The notions of being old, mature and unsophisticated have been associated with the automotive industries, and wrongly is the conclusion of this essay. It is difficult to develop and manufacture reliable, safe and fun to drive automobiles profitably under low wage competition from firms in developing economies, which immediately copy every novel idea into their own products. But automobiles with increasingly sophisticated features will always be demanded in increasing numbers by able-to-pay customers in the industrial world. Able automotive manufacturers should therefore find it profitable to supply variety in product characteristics, creating comfortable, reliable, safe and fun to drive vehicles under widely varied circumstances, and also offering all the desired status characteristics, such as unique styling, that richer customers are willing to pay for. The upper end of automotive manufacturing has become a fashion industry.

The global automotive scene

The automotive industry was once the star industry of the future, and Klepper (2006) tells the story of the enormous proliferation of car manufacturers in the US between 1895 and 1966, which was followed by a massive exit flow, as inferior manufacturers were competed out of business. This finally lead to a concentration around Motown Detroit with only GM, Ford and Chrysler left, but supplied by a US-wide component and specialized subcontractor industry. The US part of Motown has now more or less collapsed, with Chrysler recently acquired by Fiat[5], GM slowly returning out of Chapter 11, and Ford recovered but radically slimmed and without its expensively acquired premium car activity with Aston Martin, Jaguar and Volvo.

Alan Mulally, brought in from Boeing in 2006 to save Ford, has taken Ford employment down to 178,000, from 300,000, introduced large scale systems integration from Boeing to make, for instance, the Focus designed in the small-car hub near Cologne in Germany on a flexible platform. Irrespective of where in the world the car is sold, on average 85 percent of parts are common to all Focus cars. This production strategy is known as *One Ford* (FT 1 December 2010: 11, *Time*, 6 September 2010: 30–33).[6]

The Italian automotive industry in the northern part of the country is composed of giant Fiat, just rescued from near bankruptcy, and a string of super sports car manufacturers (Ferrari, Lamborghini, Maserati, etc.), most of them owned by Fiat.

The Japanese automotive industry rapidly copied European and US technology during the 1970s and wiped out an inattentive British automotive industry through price competition in the process. Japanese cars are not particularly inexpensive today and not much fun to drive, but they are durable and low on the defects that have given the US car industry a bad reputation. The Japanese manufacturers, however, now face stiff price competition from Korean, and soon Chinese upstarts, and are scrambling to be able to offer innovative products in the German quality range. Japanese car manufacturers that once taught Europe and the US automotive firms effective manufacturing, are currently "learning quality manufacturing back" from Europe (*Ny Teknik* Nr 49 3 December 2003: 15).

The UK standard car industry did disappear but is still around as a successful and leading developer of racing and sport cars, and an automotive engineering consulting industry, together employing some 50,000 people around 2000. In addition a new industry of specialized automotive subcontractors serving the global market and employing some 50,000 people has emerged (DI 13 December 1999: 17, 3 June 2005: 18f.).

The UK still has some special manufacturers of almost handmade luxury cars such as Aston Martin, Bentley (owned by VW), Rolls Royce (owned by BMW) and Morgan, most of them foreign-owned. Indian Tata has acquired Jaguar and Land Rover, but keeps manufacturing cars in England, and has been investing heavily in England to compete in the prestige markets (*The Economist*, 10 September 2011: 36f; IHT 29 May 2013). Most probably the total factor productivity of these small scale businesses, if correctly measured, is higher than it would have been in the old and obsolete British motor car industry, had it been saved. Such specialized operators also often have a higher profitability, when correctly calculated. And in the midst of distress new ventures are started. The McLaren F1 racing team announces (FT 19 March 2010) plans to build 4,000 high-performance sports cars in the UK in 2014 in competition with Italian Ferrari and Lamborghini, and German Porsche. British-based, but not British-owned, special car makers have had a surprisingly strong showing at the 2011 Frankfurt auto show writes International Herald Tribune (IHT 12 September 2011: 18) with new Aston Martin, Bentley, Jaguar, Lotus and Mini models. Tata (India) owned Land Rover also launched its mini SUV Evoque in September 2011 (*The Economist* 10 September 2011: 36).

So, recent history has witnessed the collapse of not only very large firms, but of entire regional automotive competence blocs, and sometimes the transformation of old structures into related, but entirely new industries. My touchdown will therefore be positive. I will argue that modern technology, notably C&C technology, offers what is needed for high wage European, North American and now also Japanese automobile producers to reorganize and refocus profitably on innovative high quality products, not necessarily to be manufactured in large volumes, to the extent they are capable of making the right choices, that is.

A precarious preoccupation with volume production

Volume-based price competition versus innovative product competition has been a strategic dividing line in automotive industry, the first strategy being focused on cost, often neglecting product quality, and the other going for smaller-scale and customized products. Since economies of scale and cost-motivated rationalization are constantly taught in undergraduate economics, and are therefore most easy to understand, that is also where the articulate visionaries are mostly found.

This is problematic in an industry where hands-on complex and competence demanding implementation is decisive for profitability. A focus on volume to be competitive with price, rather than with product innovation, has also contributed to a constant overcapacity at the low end of automotive markets. Scale is, however, a different concept today than on the Ford manufacturing lines that once made one standardized car affordable to the below average US income earner.

Today, there is also the large and increasing initial R&D investment associated with each new *automotive platform.* Firms have to distribute that *platform investment* over increasingly larger volumes, or through higher price mark ups. The latter is only possible if you have a unique product that appeals to the upscale customer, and uniqueness has to be achieved at competitive costs. Developing a common technical platform for a large number of models with different features is the way to achieve the first. The notion of the "world car" based on such a platform is easy to understand. This ambition, as mentioned, has been around for decades and mostly resulted in failure.

A *second* form of volume production is to be a *complete automotive manufacturer* offering a wide range of models, to satisfy every conceivable customer. Many large automotive manufacturers in fact operate across the whole range of transport equipment. Some argue that there are synergies, for instance between aircraft and automobile technology. This has probably been true to some extent, as the case of Saab Automobile demonstrates (Eliasson 2010), but on the scale tried by Daimler Benz, under its visionary CEO Eduard Reuther during the 1980s, it was a close disaster[7] (see below). And two attempted mergers between Volvo Truck and Scania revealed that there were few synergies between trucks and automobiles even at the subcontractor levels. Even so, Volvo's ambition, after having sold Volvo Car to Ford in 1999, was said to be to become "the world's leading supplier of commercial transport solutions."[8]

A *third* form of volume production has been *risk diversification* into a range of more or less unrelated markets to insure against cyclical variations in the main market. Thus, for instance, Volvo invested in oil exploration in the 1970s and 1980s, at great losses, to ensure a steady demand flow growth. Also, modern economic theory has a strong intellectual bias towards emphasizing economies of scale and risk diversion because new technology enters economic models as cost saving improvements and/or risk spreading. This theory converted into a business model is logically simple and intellectually convincing, even if wrong. Since formalized business calculation draws directly on the standard static economic model the contributions to misunderstanding are great even at the business level. And the truly experimental character of business is totally lost sight of, or worse, treated as insurable risks, a concept Knight (1921) regarded as nonsense in a business context. When most of the already giant automotive producers in the world go for scale, complete supply range and/or risk diversification into unrelated markets, the problems not only reach national political levels (employment concerns), there will also be a permanent oversupply and price pressure at the low end of the market.

Moving up the value chain

Moving up the value chain and closer to the high-income customers is difficult. It requires the art of developing and marketing unique product quality and variety of features to a small number of geographically dispersed customers. Since this has made the product development and marketing investment a rapidly increasing share in the total cost structure of such products, it has also become critical to get the organization of product development both right, and staffed with economically aware, technically innovative, and cost effective engineers.[9] This is typical of advanced engineering, and the automotive industry is no exception. To come closer to the paying customer also demands the ability to not only develop the desired characteristics, but also some distinguishing characteristics of the vehicle. Can Porsche keep its distinguishing image with gigantic VW as the owner? asks *International Herald Tribune* (28–29 August 2010).

Swedish Volvo Car and Saab Automobile were pioneers in making passenger safety, originating in military aircraft industry, a selling argument in their total automobile concept (Eliasson 2010: 150ff). Even though a local, safety-oriented subcontracting industry developed as a consequence (for instance Autoliv, the world´s largest automotive safety firm), the two firms can no longer claim to offer the safest cars in the world and have also lost important elements of their previous whole product competence by having been made into integrated parts of troubled US Ford and GM for many years. Volvo Car was sold on to Chinese automaker Geely in 2010 to fill a quality hole in that company's product line. Volvo's heart will however remain in Sweden, writes *Financial Times*[10] (29 March 2010: 16). Saab Automobile has been through a number of messy ownership and financial rounds, including a brief ownership spell in 2010 with Dutch sports car maker Spyker, and bankruptcy in 2011 to be acquired in 2012 by Japanese–Chinese electric car group NEVS (DI 14 June 2012: left).

Volvo made an attempt in the 1990s to enter the luxury car market. It teamed up with the UK based TWR company operated by racing car driver Tom Walkinshaw to form AutoNova in 1995 to make Volvo cabs, and to "generate more excitement around the staid Volvo brand". TWR boasted to be the only engineering cosultancy, besides leading German Porsche, capable of designing and manufacturing a whole vehicle, and aimed to move "to the production lane" (*Financial Times*, 29–30 March 1997: 22), to test out its new smaller-scale production techniques in Volvo's Uddevalla plant[11]. Tom Walkinshaw thought there was "a lot of needless mechanization in the motor business" and believed his new techniques to "be valid for up to 100,000 units a year".

The Uddevalla plant at the time was said to be revolutionary in replacing the production line with a permanent workstation where a group of workers built the whole car. It did not work. A conflict with TWR terminated AutoNova in the early 2000s and the standard production line was reintroduced in the Uddevalla factory in 2002. Italian Pininfarina entered as a partner in 2003. Volvo cabs were not a business failure, but were close to being so[12].

The critical market for specialized subcontractors

With innovative product variation increasingly demanded by upscale customers, distributed production becomes much more than outsourcing the low end of manufacturing. As long before in aircraft industry, engineering industry is becoming increasingly dependent on developers of sophisticated subsystems, and on the existence of a broad market supply of such specialized subcontractors (see further companion paper, Eliasson 2011). Such distributed production has also become key to achieve both economies of scale through networking externalities and flexible product variation in response to customer demands. Not only do the firms at the upper end of the advanced product line facing the final customer have to be good at that; subcontractors also have to understand how to fit into the whole to be allowed to participate (Eliasson 2010: 135ff.). A number of automotive firms in the world have found this difficult, and apparently many volume producers have not been able to change their ways.

Large-scale systems integration means concentrating on product development, outsourcing non-core physical manufacturing on specialized subcontractors, and then marketing and distributing the product, sometimes even taking over part of the maintenance and servicing of the product from the customer. This technology was developed in aircraft industry and Alan Mulally has made a point of having brought it with him to Ford in 2006 from Boeing (*Time* 6 September 2010: 30f). This is part of a general movement *to come closer to the final customer, and to be responsive to rapid and unpredictable changes in customer demand.* However, the preoccupation with standardization and volumes has permeated the procurement functions of the large automotive companies with consequences for the specialist subcontracting industry. *Covisint*, founded by GM, Ford and Daimler Chrysler 2000, has developed into the world's largest Internet market for components and subsystems in the industry. One silent ambition was to cut prices for components through competitive purchasing in more transparent markets (the perceived, great advantage of Internet trade, McKnight and Bailey 1997), but the official reason for establishing this trading place was to facilitate the development of new organizational solutions for production over the markets of subcontractors. VW and BMW, however, decided not to participate. They did not want to become dependent on US standards, and this observation is a point of departure for my final discussion of the problems of far reaching standardization and rationalization through volume expansion and cost cutting, on the one hand, and advanced innovative product development, customization and smaller scale on the other.

One should recall here that one reason given for the Daimler Chrysler merger failure and break up in 2007 was that the sharing of parts between Chrysler and Mercedes, even though seemingly a logical advantage, did not work. Mercedes customers did not appreciate standardized and cheap Chrysler components, while Chrysler customers did not want to pay for expensive Mercedes parts.

While C&C technologies have made it possible to handle increased product complexity and flexibility in both product development and manufacturing organization, the same technologies have also made it possible for managements to deal with the same complexity, and C&C technologies have facilitated the corresponding manufacturing reorganization at all levels. Most important, as elaborated in the companion paper Eliasson 2011, *integrated production has made it possible for individual producers to operate on a small scale and still enjoy economies of scale through networks of specialized subcontractors*. But success has required the learning of new management competences. A particularly *important contrast* has therefore been played out between *competition through innovative technological product development*, on the one hand, and through *the building of economies of scale to be able to compete with volume manufacturing, cost reductions and lower prices on standard products,* on the other.

3 The European automotive scene

European automotive industries exhibit the whole range of experimental variety, and Europe is crowded – some would say overcrowded – with automotive factories manufacturing some 15 million autos year. Strategic management volume minds are pitted against the experience of small innovative product developers. Automotive manufacturers have been pushed by both internal and external competition and a global overcapacity to reorganize their ways. Some "experts" said in 2004 (DN 15 October) that German manufacturers must cut costs dramatically to prevent production from being moved to East Europe and Asia. The large part of the UK owned automotive industry in the UK was wiped out in the 1970s. Today Indian Tata is England's biggest manufacturer with 19,000 (of a total of 45,000) workers in Jaguar, Land Rover and engine plants (*The Economist* 10 September 2011: 36). BMW still successfully operates Mini and Rolls Royce production facilities in the UK, but both products and facilities have been completely upgraded to German standards. What remains is US and Japanese owned and operated. In time for the 2011 Frankfurt auto show British-based, but not British-owned car makers have made "a surprisingly strong showing" with new Aston Martin, Bentley, Jaguar, Land Rover, Lotus and Mini models (IHT 12 September 2011: 18). On the whole, writes Sweden's *Dagens Industri* (29 and 30 November 2012) the British automotive industry that disappeared in the 1970s, including government operated closing British Leyland, is now coming back strong employing a total of some 700,000 in Great Britain, mostly in small-scale specialised automating and engineering subcontractors.

The bankruptcy of century-old Italian Fiat (founded in 1899) was a close call in the 1990s. Volume was, and still is, the key strategic concept for Fiat. Sergio Marcionne, CEO of the company in previous severe distress, has turned it around brutally, observing that the 2009 meltdown in automotive markets offered great opportunities for cheap acquisitions, too tempting to pass. So he offered to buy ailing Chrysler, from Daimler Chrysler, and Opel from GM. Daimler Benz had failed to make something of its acquisition from 1998 of Chrysler at the tune of billions of Euros (see below). Marcionne says: "There will be only five or six automotive groups in the world in a few years. The minimum for a volume maker competing in every sector is about 5.5 million cars" (*The Economist,* 9 May 2009, DI 29 16 January 2010: 16, SvD 31 January and 29 September 2010). So it is not surprising that gigantism features prominently in the automotive strategy repertoire. And it is not clear that Fiat in the end will belong to the exclusive group of surviving volume producers.

Let me go through some stories from the *experimental history of recent European automotive industry,* comparing the experiences of Germany, notably Bavaria/Baden-Württemberg (B/B-W), with those of Sweden, notably west Sweden, and the Lake Mälar district around Stockholm. The role of the local subcontracting industry will be specially addressed.

The B/B-W area is significantly larger in all dimensions than the two Swedish automotive regions combined, in which transport industry is located (The Lake Mälar and South West regions), except in geographical size. B/B-W have 22 million inhabitants, most of whom can commute daily into a common labor market. In that region we find Audi (the luxury sideline of VW), BMW, Daimler Benz and Porsche. The region has the world's perhaps most developed and sophisticated range of specialized automotive industry subcontractors, with Bosch in the centre. An industrial culture has developed that fosters a unique balance between competition and cooperation (Eliasson and Eliasson 2005b: 426ff.). Outside the B/B-W region we find VW headquartered in Wolfsburg, Nieder-Sachsen, still GM owned Opel in Bochum and Rüsselsheim, Ford's global small car hub near Cologne, and automotive and subcontractor plants scattered all over Germany.

The geographically much larger Swedish area includes two concentrations with together some four million inhabitants: one on the West Coast (Volvo, Volvo Car and Saab Automobile) and the Lake Mälar region (Scania and Volvo Construction Equipment, VCE). Perhaps also South East Sweden with Saab, the aircraft manufacturer in Linköping should be included. Far away in the North (Örnsköldsvik) we find the specialized military vehicle producer Hägglunds, now owned by British BAE systems. None of the Swedish automotive concentrations are within daily commuting distance of one another.

3.1 Southern German automotive industry

Daimler Benz (with its CHQ in Stuttgart) was founded in 1926 when Daimler merged with Benz and Ciel.[13] In the 1980s its chairman Eduard Reuther tried to develop a *technology power house,* with aircraft industry in its core, to feed automotive industries, and including white-ware manufacturer AEG, with sophisticated technology. The strategy looked principally OK on the surface, but failed miserably in practice. The entire business group, however, was supported financially by profits from the Mercedes luxury car division. Reuter's successor Jürgen Schremp transformed Daimler Benz from a money-losing mess, as Business Week expressed it (7 August 2000: 19) into a solid profit-maker between 1995 and 1998, then aimed for another tack and went for scale with the strategy to develop and manufacture a complete range of cars, from the Smart mini car, over the A series to the big S sedans, and on to heavy trucks. To support that, failing US Chrysler (the innovative US automobile star manufacturer of the 1950s) was acquired, and Daimler Chrysler was formed in 1998. The need to offer a full range of car models and economies of scale were quoted as the strategic reasons for this gigantic approach. The US lead in virtual design technology[14] was another quoted reason for the Chrysler acquisition. This strategy almost wrecked the company. It again survived thanks to profits from the luxury car division, which had, however, begun to slowly suffer from insufficient attention (BW 7 August 2000). Schremp had to quietly leave through the back door in 2005 (DI 19 December 2005: 14), the merger was dissolved in 2007 and Chrysler was picked up by Fiat in 2009.

BMW (with CHQ in Munich, founded in 1916 to make aircraft propellers[15]) was much smaller than Mercedez Benz as an automobile manufacturer in the early 1970s. BMW, however, opted for an entirely different strategy. It stayed with its last in the 1980s, and focused on small, luxury and sporty cars that were both enjoyable to drive and useful. After a silent agreement had been broken by Mercedes in 1983, when Mercedes introduced the "baby Benz", or the 190 series (BW 9 December 1985), that BMW should concentrate on small cars and Mercedes on large cars, intense competition between the two followed. BMW entered the large car market, but stayed with luxury models. With the expensive sidestep in 1998, when it acquired hopeless British Rover (FT 18 November 1998), BMW managed to expand profitably into the largest luxury car manufacturer in the world, selling better and more enjoyable to drive cars at a significant, but not excessive margin compared to hopeful competitors. The failed Rover venture was written off in 1998 to the tune of billions of Euros.[16] BMW kept Rover's Mini brand and has succeeded in developing that into a profitable small and fun to drive car under a separate brand. The Mini is manufactured in the UK (FT 11 November 1998).

The logic behind the merger mania decisions had been clear and intellectually convincing, even though wrong, as a strategic business concept. In 2011 BMW was the second most recognized and valuable auto brand at $22.4 billion, after Toyota's $24.2 billion, and well ahead of the third player Mercedes' $15.3 billion (SvD Näringsliv 6 September 2011: 22), and claimed proudly that it would not make a loss if a repeat of the 2008/09 crisis would occur (FT 10–11 September 2011: 8).

Porsche is the smallest of the German automobile makers (with its CHQ in Stuttgart. Founded as an engineering consulting firm in 1931[17]), but enjoys the highest profit margins in the automobile world. Porsche management claimed that it could develop and manufacture "at prices larger companies cannot achieve" (BW 11 December 2003) and the solution is "an efficient manufacturing base that outsources 15 to 20% of production" combined with a hands on hardware cost control. Even though it is a "volume producer" in the sports car market – compared to, for instance, Italian Ferrari – Porsche has kept the number of cars sold small. It has, however, diversified away from its narrow base in series production of high performance sports cars to include also a "sporty" SUV and a "family" car, and has recently introduced a environmentally correct hybrid luxury car. True to its tradition, Porsche also operates one of the world's most respected engineering consultancies.

During 2007 Porsche used its considerable financial resources to acquire a controlling stake of 31 percent in VW, a possibility that opened up when the EU Court threw out the special law that protected VW from hostile takeovers. VW thereby became a subsidiary of the family firm Porsche (controlled 50/50 by the families Piech and Porsche) which, for a while, seemed close to controlling the €150 billion automotive empire of VW, Porsche, MAN and perhaps soon Swedish Scania combined (FT 4 March 2008: 13,23).. With a voting stake of 30 percent in VW, Porsche's CEO declared that "we are of course interested in exercising our influence fully (DI and Sv D. 24 October 2007; DI 25 February 2008).

In 2008, however, funny things began to happen at Porsche. Its innovative treasurer had managed to corner the market for VW shares and earned paper capital gains larger than Porsche sales that drove up VW stock prices, and a handful of hedge funds that thought Porsche was about to acquire VW (and perhaps rightly so) into bankruptcy, or nearly so. Luck, however, ran out for Porsche. In the end the company had accumulated debt of some 100 billion SEK, after the raids on VW during 2007 and 2008. Banks became reluctant to support that debt mountain and Porsche's advances for support were rejected by the German state. So while Porsche SE (the parent and 51 percent owner of VW) was in a financial mess, Porsche AG was still the most profitable automobile maker in the world (SvD 18 March 2010). In the end VW (and von Piech) lent €750 million to its owner Porsche SE, which had to sell 49.9 percent of Porsche A6 to VW (DI 2 July 2009; FT 4 March 2008). The VW/Porsche relationship had been turned upside down, with VW "acquiring" Porsche. Von Piech in passing also managed to get rid of the small scale profit oriented and highly regarded CEO Wendelin Wiedeking of Porsche.

In September 2011 VW announced that it would not complete its merger with Porsche because of the unclarified legal risks associated with the pending lawsuits filed by the loss making institutional investors, alleging market manipulation, but changed its mind in 2012 to exploit a tax loophole and made Porsche a part of VW, then added that it might still buy the half of Porsche's car business that it did not own (FT 10/11 September 2011: 11; 14 September 2011: 17; SvD, N 6 July 2012: 19).

VW (Headquartered in Wolfsburg in NiederSachsen, outside the B/B-W region. Founded in 1938) in reality now owns the financially troubled Porsche SE. Von Piech, chairman and large owner in Porsche, has earlier succeeded in reorganizing and expanding VW into something new, building layers of brands, from inexpensive Skodas, via the VW semi luxury brand range of cars to the **Audi** (Headquartered in Ingolstadt, Bavaria. Founded in 1909. Acquired by Volkswagen in 1965), the latter having been carefully cultured (by its then CEO von Piech) into a luxury brand that competed directly with BMW and Mercedes.

Audi, in fact, in 2010 was challenging BMW as the largest maker of luxury cars in the world. Audi is separately managed from its CHQ in Ingolstadt in Bavaria. At the upper end VW has also acquired Italian Lamborghini and the Bugatti brand name, and developed the fastest series-built sports car ever that tops 400 km/hour.[18] But VW top management ambitions do not end there. VW did not have its own large truck division, but only a range of smaller transport vehicles, so it has been eyeing both MAN (also a Bavarian company with CHQ in Munich) and Swedish Scania, and acquired a 34 percent voting share bloc in Scania in 2000, and a 15 percent stake in MAN in 2006. The former was only possible because the dominant owner, Swedish Investor, for some very peculiar reason was willing to sell (see below). On the Scania acquisition, however, *Financial Times* notes (on 14 September 2011: 17; DI 16 September 2011: 25) that VW's "hopes of global domination" must await regulatory clearance "in the European Commission", a clearance that was granted shortly thereafter. Von Piech, apparently has not allowed himself to be restrained by the merger debacles of Daimler Chrysler and BMW, and appears to entertain ambitions to build the world´s largest and most complete automotive business, including everything from small and inexpensive Skodas to luxury Bentleys via super performing Bugattis to heavy trucks. VW announces its decision to invest some €45 billion (570 SEK) over the next five years (DN 17 September 2011: 3). Is this also a gigantic experiment in megalomania? Daimler Chrysler failed. Is VW the next candidate of comparable scale, once boldly driving von Piech has to turn the management task over to a less talented and less forceful successor?

Key to the successful automobile competence bloc in Southern Germany is its outstanding and close market of specialized subcontractors with innovative **Bosch** (headquartered in Stuttgart) at its core, and a culture that combines competition with cooperation in a constructive way. One example is a development project of BMW subcontracted to Bosch. My argument (during one of my interviews) that a BMW development project at Bosch will also benefit Mercedes and Audi was brushed aside by the assertion that next time BMW will benefit from a Mercedes project. Another story has it that when Audi failed to develop an aluminum V8 engine for its luxury car line, BMW and Mercedes stepped in to help Audi get its V8 in shape. The reputation of the entire German luxury automotive industry was at stake.

3.2 Swedish automotive industry

Also, the *Swedish automotive industry* has been subjected to indigenous political visions of its own. Volvo, Saab and Scania[19] have been innovative players that have helped make Sweden a prominent automotive manufacturing economy. The repeat story has been that all three companies have been too small, too dependent on a narrow product line and on unpredictable customers and business cycles to manage on their own. For instance, after a failed attempt to merge the Volvo Truck division and Scania in 1977, which was stopped by stubborn resistance from Scania management, the CEO of Volvo, Pehr Gyllenhammar, changed his vision and attempted to forge an engineering/oil deal between Volvo and Norway, a diversification to spread business risks. That attempt also failed, this time because of Volvo owner resistance.

Today, however, the Swedish automotive industry is in a shambles, Volvo having been acquired first by Ford in 1999, only to be sold on to Chinese Geely in 2010, and Saab Automobile first to GM in two stages 1990 and 2000, only to be sold on, after a messy period of negotiations, to Dutch sportscar maker Spyker in 2010, and then in November 2011 being close to bankruptcy for a year, after negotiations with Chinese Pang Da and Youngmanhad had failed because GM would not allow the Chinese to use technologies embodied in the current Saab models.[20] Two Swedish studies (from Tillväxtverket and Tillväxtanalys) announced that Swedish auto making is too small, uncompetitive and has no future (SvD 13 May 2009: 4f). The two heavy truck manufacturers Volvo and Scania, and Volvo's construction equipment business (VCE), however, still belong to the most competitive businesses in its markets in the world, with the slight distinction that Scania is about to be swallowed up by VW/MAN.

Let us look back to see how that has happened.

Volvo (with CHQ in Gothenbourg) was founded in 1926 as a spin off from the Swedish ball bearing company SKF. The first Volvo automobile rolled off the production line in 1927. Volvo gradually emerged as a quality producer of durable, affordable but not fancy cars that could be sold in export markets at reasonable prices, at the time because of low Swedish wages. The "responsible" minister of trade, however, laughed at Volvo management when it declared its ambition to enter the US market in the late 1950s, and told the exchange control authorities to block the venture. Volvo, however, persisted and eventually got the needed permission (or evaded the regulation) to move money to the US, and established itself very successfully there. By the early 1970s the US was Volvo's largest market.

The first Volvo Truck was successfully introduced already in 1928 and became a success. The production of buses was a natural consequence, commencing in 1935. Aircraft engines were added to the widening product repertoire in 1941, when Volvo acquired Nohab Flygmotor[21] as demand for military aircraft increased rapidly (Eliasson 2010: 90ff.).

Boat engines (the Penta company) were acquired in 1934 and with the acquisition of Bolinder Munktell in 1950 Volvo entered the markets for both agricultural and construction equipment. The latter acquisition was later complemented with Åkerman excavators, and eventually the whole business was streamlined into what is now Volvo Construction Equipment (VCE) in Eskilstuna, the third largest company of its kind in the world. The Volvo Group today is apparently a quite diversified conglomerate which has difficulties internally capturing technical synergies. This is so even within the group's individual companies. To illustrate the key theme of this paper, in October 2011 VCE announced that customized heavy construction equipment developed and manufactured by the specialist Malmö Ce De Group will be sold under the Volvo CE brand. This is the first time Volvo includes externally manufactured products in its own product line up (DI 31 October 2011: 22).

In 1972 Volvo acquired a dominant position in the Dutch automotive manufacturer DAF, partly to broaden its car range into small cars, partly to access the believed to be unique and important continuous gear system developed by DAF. Volvo soon became an innovative product developer both in automobiles, heavy trucks and in construction equipment. Safety and durability in the cold Swedish climate became guidelines for the development of Volvo automobiles from the beginning. As it happened, Volvo Aero hydraulic technology from military aircraft engine development helped Volvo to successfully pioneer hydraulic technology in its trucks and construction equipment (Eliasson 2010: 95).

During the years 1975–1976 Volvo Car went through a period of serious quality problems and lost a large part of its US sales. Financial troubles followed. As mentioned, a merger between Volvo and the Scania division of Saab Scania to exploit synergies was discussed at the upper ownership levels in 1977. It did not lead anywhere because of stubborn resistance from Saab Scania management.

The attempt of Mr Gyllenhammar to "merge Volvo with Norway" resulted in the Norway Agreement ("Norge Avtalet") in the late 1970s. The business strategy then was risk diversification, another form of economies of scale. A shares issue directed at Norwegian interests was planned. Norway would obtain Swedish industrial assets and technologies in return for oil concessions in the North Sea. Volvo share owners, however, objected, and the "agreement" came to nothing.

So far Volvo growth had, to a large extent, been organic, but attempts to reach scale and financial stability as a hedge against unstable automotive demand continued. In 1981 Volvo acquired Beijerinvest and with that came a diversified portfolio of oil (STC) and food (Provendor) and pharmaceutical assets. Volvo became a large owner of Pharmacia that merged with Government owned Procordia into which Volvo placed its food assets. In the mid 1980s the US truck manufacturer White was acquired.

A new attempt to gain size and financial stability came in 1990 in the form of an alliance with French Renault and mutual minority ownership. In 1993 a merger with Renault was planned, but came to nothing when Swedish share-owners blocked the idea to become dependent on the French state through its so-called golden share in Volvo, negotiated behind their backs by Mr Gyllenhammar. The alliance was dissolved and CEO Gyllenhammar was forced to resign.

A rapid expansion through acquisitions now followed. In 2001 Volvo Truck acquired Renault Truck and Renault´s US subsidiary MACK, paying Renault with its own shares (20 percent). Renault now was Volvo´s largest owner. Renault continued to buy Volvo shares in the market (Eliasson & Eliasson 2005b: 423). The Japanese truck manufacturer Nissan Diesel was acquired 2006 and 2007, and the Ingersoll Rand construction equipment division in 2007. With the Nissan Diesel acquisition Volvo Truck had acquired a particularly effective super capacity hybrid technology, superior to battery technology in stop and go city traffic (*Ny Teknik*, 28 February 2007, Nr 9). The ambition of Volvo was now to become the world's leading supplier of commercial transport solutions.

Fragmenting the Volvo Group to spread risks and to stabilize demand, however, also meant that management resources were stretched thin and attention to the automotive business suffered. Volvo Car, being part of the fragmented Volvo Group of trucks, buses, construction equipment, boat engines and aircraft engines, had been reasonably profitable, but in the end the group was found to lack sufficient financial and management resources to stay competitive in all these markets, and in particular in the crowded automobile market. It was considered necessary to increase both scale and resource access. A merger with giant US Ford was seen as a solution for Volvo Car in 1999. Volvo Car was placed in Ford´s premium car line up together with Aston Martin, Jaguar and Land Rover. Ford, however, was already bigger than the protective haven envisioned[22] and part of the chronic decline of US automotive industry. Much publicized attempts to convert its manufacturing lines to flexible production and modular platforms with common modules and interfaces to raise volumes and cut costs did not help (FT 11 November 2002: 19, 1 December 2010: 11). Ford was soon forced to sell its premium car activity to save itself. Volvo went to Chinese Geely in March 2010, with little to show for itself except a need of a premium car (FT 29 March 2010: 16).

In 1999 Volvo made another attempt to take over Scania when buying 13 percent of Scania shares in the market. This time the Wallenberg and Volvo owners appear to have been in agreement. Investor sold 39 percent of its Scania shares, and purchased 13 percent of Volvo voting shares "in return" (BW 30 August 1999: 24, *Affärsvärlden* 19 March 2008).

The EU, however, promptly blocked the merger attempt in March 2000 and Volvo got stuck with a large block of Scania shares (45 percent of capital and 31 percent of votes). Investor, as mentioned, sold 18.7 percent of its Scania capital (34 percent of votes) to VW. This was at the time of the "New Economy" discussion, and Investor was trying to recompose its asset portfolio away from its previous dominance of "mature" engineering assets.

The EU now demanded that Volvo sell its shares, and it finally and reluctantly did so through a separately created holding company: Ainax. In 2004 Scania (through Investor, which apparently had changed its mind) acquired shares in Ainax, and finally the whole company.

The ownership mess did not end here. In 2002 Scania, and the private equity arm of SEB (EQT), had considered buying MAN. MAN and Scania began to cooperate in developing gear boxes. In 2006, however, MAN placed a hostile bid on Scania, and VW acquired 15 percent of MAN (DI and DN 15 September 2006). The offer was rejected. MAN acquired more Scania shares in the market and VW increased its ownership in MAN to 30 percent. In 2007 VW bought more Scania shares and suddenly (together with MAN) owned more than 50 percent. Finally Investor and the Wallenberg Foundations sold all their remaining Scania shares to VW (*Affärsvärlden* 19 March 2008). According to one story, the sale of Scania is the greatest-ever blunder of the Wallenberg family. It began in 1998 when the new chairman of Investor (Percy Barnevik) formulated the new strategy to leave the mature engineering industry and enter the New Industries of the New Economy (SvD 4 April 2010).

It now looks as if VW/MAN are ready not only to take over Scania, but also to steamroller the recommendations of the Swedish Board Nomination Committe and the large number of remaining Swedish share owners in Scania (DI and SvD 1 April 2010). And the Swedish ownership mess is still not over. When Renault engages in an alliance with Volvo's worst competitor in the truck market Daimler, other shareowners demand that Renault leave the Volvo board (SvD 31 March 2010). Renault is shifting loyalties and is soon prepared to sell its Volvo block of shares, and finally does to temporarily weather its own crisis in the car market (SvD 1 April 2010, 14 December 2012, DI 21 June 2010: 6f.). An alliance between Renault, Nissan and Daimler was announced. Cooperation is needed to survive the recession the Renault CEO Carlos Ghosn announces (DI and SvD 8 April 2010).

Scania was founded in 1891 and is headquartered in Södertälje outside Stockholm. It has long been one of the three global leaders in the market for heavy trucks. It is much smaller than Volvo Truck, Daimler Truck and smaller than Iveco and MAN, but the by far most profitable. It has followed an internal growth strategy. Since 1948 Scania has also been the Swedish agent for VW automobiles. The strategy argument of Scania has been that economies of scale can be captured by other means than by entering markets everywhere through acquisitions to raise volume (*Ny Teknik* Nr. 8. 18 February 2004). Its key to success has been *increasing returns to product development through ingenious ideas and competent engineering*. With the right modular design Scania has achieved a very large variation in (customization of) its products. Combined with the "Toyota lean way", but without the hierarchical Japanese leadership, Scania developed its successful organization to achieve economies of scale without overextending its size, says CEO Leif Östling in *Ny Teknik* (16 December, Nr. 51–52).[23]

The two Swedish heavy truck manufacturers have therefore pursued very different strategies. While Volvo spread its resources thinner aiming for volume and being everywhere, Scania did nothing of the kind. Scania had focused on what it was best at, and refrained from being elsewhere. Thus, for instance, it backed out of an attempt to enter the complicated US heavy truck market. Scania has been the most profitable producer of the two. Competition between the two Scandinavian heavy truck manufacturers is also believed to have contributed positively to their respective performance (SvD 8 July 2003, Eliasson and Eliasson 2005b: 447). The two failed attempts to merge Scania and Volvo Truck, however, left Scania in the precarious ownership situation that may force it under the rule of giant VW.

The VW/Scania/MAN affair is interesting (*Affärsvärlden* Nr. 12–13, 19 March 2008, and Eliasson and Eliasson 2005b). MAN had been an ailing truck manufacturer until Håkan Samuelson was recruited from Scania in 2000 to turn it around (SvD September 14. 2006), exporting his experience of superior Scania product and manufacturing technology. He soon, however, fell into disfavor at some higher MAN or VW level, and after some management reshuffling the attempts to integrate Scania and MAN were managed by a VW executive, and VW removed several Swedes from the Scania board (DI 1 April 2010: 8). On this, Leif Östling had previously announced that potential synergy gains from integrating Scania and MAN are much smaller than they are believed to be, at least for Scania.[24] He adds that it was not even possible to find economies of scale in a common development of gear boxes 2005 (DI, interview 2009: 18). On 15 November 2010 Scania confirmed that it is investigating the possibilities of increased cooperation with MAN. On 5 May 2011 CEO Leif Östling's contract was prolonged for three years. On 9 May 2011 VW announced that it had increased its share of MAN stock to above 30 percent, and again on 28 June. VW now owns 55.9 percent of MAN. Pressure was placed on Leif Östling to initiate cooperation with MAN (DI 14 July 2011: 26). In June, however, the European Commission had already told VW that it "must await regulatory clearance before pursuing closer cooperation between MAN and Scania, setting back its hopes for global dominance". This clearance, however, had already been obtained on 26 September (FT 14 September 2011: 17; DI 27 September 2011: 19). And so Leif Östling is suddenly nominated to the executive board of VW and the merger of Scania and MAN is postponed (DI 2 and 4 June 2012, SvD 25 April 2012).

Saab Automobile was founded in 1946 as a division within Saab, and as part of Saab's post-war diversification program into civilian markets. The first two-stroke Saab 92 was introduced in 1947, and series manufacturing for the market began in 1949. Saab Automobile (established as a separate company in 1990 with its CHQ in Trollhättan) has been a loss-maker for most of its life. It has, however, been technologically innovative. The Saabs were sporty and fun to drive compared to the Volvos. Saab pioneered a number of safety features that it learned from Saab Aircraft. The first side-impact protection system in the world was introduced in 1972 when Saab aircraft engineers simulated the optimal protective grid structure for the doors. Saab Automobile, however, could not develop the sporty, luxury and affordable family brand that was considered a possibility (Eliasson and Eliasson 2005b: 424f.). It was sold to giant US GM in two stages – 1990 and 2000. GM needed a quality brand and was believed to be able to offer the financial and market support Saab Automobile lacked. GM was already in serious trouble by 2004, and on its sixth straight loss year. With giant GM in Chapter 11 the fates of its subsidiary companies, Saab Automobile and Opel, were increasingly uncertain. GM considered shutting down either its large Opel plant in Rüsselsheim in Germany, or Saab Automobile (DN 15 October 2004: 6). Employment dominated the political discussion in both Sweden and Germany. Just recovered (from near bankruptcy), Fiat offered to buy Opel, but was rejected by GM in favour of the Canadian engineering firm Magna, a deal supported by German politicians, only to be thrown out in late 2009, when GM decided to keep Opel.

4 Will there still be a Swedish automotive industry after this market mess?

With both Volvo Car and Saab Automobile put up for sale to save the US parents financially, the fate of the entire Swedish automotive industry was at stake. Most Swedish policy discussion, unfortunately, was wasted on the unemployment concerns of Swedish politicians. More important, and not much discussed, have been the loss of systems integration competence to develop a complete and sophisticated automobile. Additional worries were that the downstream, large, and in places very competitive and specialized Swedish automotive subcontracting industry (for instance automotive safety firm Autoliv and four wheel drive developers Haldex[25] and Köpings Mekaniska Verkstad[26]) would find it in their commercial interest to move abroad. Autoliv and Haldex have both been considered acquisition objects for years (DI 13 and 17 November 2004). The hydraulic four-wheel system of Haldex was also sold to US Borg Warner in 2011. Only about 5 percent of Autoliv manufacturing, employment, sales and ownership are in Sweden, even though the CHQ is still in Stockholm. Less than 25 percent of the Volvo automobile is in fact Swedish (*Ny Teknik* 3 December 2008, Nr. 49).

Saab Automobile had been courted by a spectrum of prospective buyers, few of which had been accepted by GM, which wanted a maximum price and minimum risk for competition in its own markets, for instance from China, with Saab technology. Tiny Swedish super-car developer Koenigsegg was considered for a while (DI 13–14 June and 17 June 2009: 6), but then rejected. Previous MAN CEO Håkan Samuelson entered with a group of Swedish investors (DI 8 January 2010: 4), only to be rejected. Finally, tiny Dutch customs sports car manufacturer Spyker won GM's favour and was "allowed" to acquire Saab Automobile (DI 25 January 2010). Spyker was ready to restart series production in late March (DI 15 March 2010). However, there were constant doubts about Spyker´s financial capacity to operate Saab Automobile in the long run, but Chinese Beijing automobile (BAIC) announced that it would be happy to enter as a partner if Saab Automobile runs into problems (DI 1 April 2010). Lack of finance, however, forced Saab Automobile to stop production. Russian financier Antonov wanted to step in, but was rejected as a financing partner by the European Investment Bank for ethical and other reasons. It was OK, however, if Antonov took over the entire loan from the bank. Bankruptcy loomed heavily over Saab Automobile. In November 2011 Chinese automotive companies Pang Da and Youngman were cleared as owners by the Swedish court that was handling bankruptcy petitions, and promised to invest 20 billion SEK over the next six years. But GM, which owns the technologies used in the current Saab models, said no (DI 1 November 2011: 8f.). Finally the Japanese–Chinese electric car group NEVS buys Saab Automobile to manufacture electric cars on the old 9–3 Platform (DI 14 June 2013: 6ff).

Volvo car had turned out a much better buy for Ford than Saab Automobile for GM. For some time Volvo was the most profitable unit within Ford (DI 21 April 2005). The acquisition price was soon returned to Ford in the form of Volvo profits. But also Ford had been unable to benefit further from both Volvo Car and the premium range of brands it had acquired (including British Jaguar and Aston Martin), and sold them back to the market at bargain prices. In the midst of the global crisis Volvo Car was therefore also put up for sale.

After a while Chinese Geely, known for uninhibited imitation and appropriation of foreign technology, became the favourite prospective buyer, and a deal was clinched in March (FT and DI 29 March 2010). Again, the possibilities of reaching the Chinese mass market through Geely were heralded as a great benefit. However, concerns have also been voiced that Geely will simply take the Volvo brand and the Volvo technology, once their engineers have learned it, and, contrary to promises given, move Volvo to China. So with Volvo Car in China, Saab Automobile also in China, or shut down, and supreme heavy truck manufacturer Scania lost to German VW in the grand business blunder of Wallenberg controlled Investor, what will be left of the Swedish automotive industry in the longer run? The Volvo Group, with Volvo Truck will remain, but what will be the rationale for Swedish automotive subcontractors to stay in Sweden, when their most competent customers may be in Germany and China?

Eliasson and Eliasson (2005b) concluded that the large Swedish manufacturing firms had been, and still were, excellent technology players, but might have lost some of its previous big business management excellence, and were now routinely pushing for volume solutions to their competition problems. Swedish businesses, furthermore, were not great commercializers of new technology, were operating in a less than excellent entrepreneurial environment, and suffered from insufficient local supplies of industrially competent finance to restructure failing large firms. However, foreign direct investors had contributed needed commercialization competence to keep Swedish manufacturing, so far, on a reasonably fast growth-path. It was even noted that the relative smallness of Swedish companies would make the competitive shake out in Sweden looming up ahead more thorough, and that Swedish industry therefore at least had a chance to come out better in the long run than Germany. The almost endless financial resources of the large German automotive manufacturers had made it possible for them to sustain giant business mistakes without thorough restructuring, and this should be considered a problem rather than an advantage. The view of 2005, however, needs some modification today.

The smallness syndrome has been a constantly returning story in the Swedish automotive discussion. Why do we need two manufacturers of both heavy trucks and passenger cars in small Sweden? Why have Swedish owners, financial markets and politicians allowed them to waste resources by competing with one another? Attempts to merge Scania and Volvo Truck, however, failed. Both companies remained globally excellent performers thereafter, even though the Volvo Group was still spreading its management resources thin on activities that contribute only marginally to each other. Scania is the smaller and the more profitable of the two. Scania, after a series of high level ownership blunders, is, however, about to be swallowed up by German competitor MAN. Volvo Car and Saab Automobile have not been as fortunate and are no longer autonomous manufacturers. Right now it looks as if both will end up being managed from China.

The automotive industry, as well as engineering at large, are still industries of the future, and the case argued some ten years ago that in the New Economy they were on their way out is dead (Eliasson 2011). Automotive, in addition, offers significant positive spillovers to other systems integrating engineering firms. The fact that competition takes place at the product level rather than at the manufacturing process level means that little can be gained in the form of cost savings by removing duplication, and the gains are insignificant compared to what is gained through intense innovative product competition between local producers. So, unfortunately, two market failures may in the end decide the fate of Swedish automotive industry, either in combination or one alone. *First*, local management competence may no longer be up to the task. Then, losing the industry, as in the UK, is nothing to worry about. On the contrary, trying to save the industry through government intervention would only blow up the problem, and prevent other superior technical, management and (now also) entrepreneurial capacities to fill in, where automotive manufacturing is leaving. It is then only to be hoped that those entrepreneurial capacities exist. However, insufficient management competence does not look to be a prime problem. Little in the previous analysis would suggest that automotive is an industry of the past for Sweden. At least one of the companies involved, Scania, has a magnificent management track record.

Second, it could however also be the case that the requisite management competence is there, but that lacklustre Swedish ownership and financial performance explains the absence of local resources to restructure the firms. This second possible market failure, which I am prepared to embrace, is the far more serious policy concern for the future, and it should rather be labelled a policy failure, since financial markets have long been half the way in the government domain in Sweden, a circumstance that has not been congenial to the fostering of the business competences needed locally to cope with the new global financial environment. So with three of the four major automotive firms being foreign-owned, with an uncertain long-term presence in Sweden, the critical local specialist subcontractors will also have to revise their long-term location strategies.

The bigness syndrome looms heavily over the global automotive industry. Attempted and completed and not very successful mergers and acquisitions characterize the market. The official reasons have varied. Most common has been the announced need to achieve sufficient volume to cover "soaring" development costs, and/or to be a *complete automotive producer*. Earlier the need to spread risks by diversifying into unrelated markets were frequently tried, and the Volvo Group is an example of that. Reducing risks and minimizing costs have been key words and all three reasons represent different ambitions to achieve economies of scale, not infrequently associated with complete failure. Daimler Chrysler is the perhaps most spectacular example.

Another quoted reason for the "volumania" syndrome in automotive industry is that large volumes are needed to cover the enormous, and rising, platform investments for every new model, even though the most successful manufacturers Porsche and Scania appear to have overcome that dilemma through increasing returns in the platform development itself.

Unexploited synergies have been quoted. Daimler Benz failed twice, even though there might have been some common technology in aircraft and automotive. Leif Östlund, still CEO of Scania, stated categorically that there were none to be gained, neither in purchasing and manufacturing, nor sales between trucks and automobiles, and surprisingly much fewer than believed between heavy truck manufacturers MAN and Scania, and even so when it came down to developing a common gear box.

Very rarely are the benefits of intense innovative product competition between local producers heard from the industry itself, even though that seems to be a main explanation of the superb performance of the Southern German automotive industry, and of Scania and Volvo Truck.

Capturing *new technology* cheaply (a business opportunity) has been a reason sometimes quoted for mergers and acquisitions. The Chinese have obviously been interested in Volvo Car for that reason (FT 29 March 2010: 16). Daimler Benz was once quoted for being interested in Chrysler′s superior virtual design technology. Fiat chairman Marcionne declared that with its Chrysler acquisition (now from Daimler Benz) Fiat will have access to Chrysler's hybrid and electric technologies and – he added – will also reach a larger volume over which to distribute total costs (SvD 31 January 2010). VW and MAN are already at work transferring Scania heavy truck know-how into their own work places.

Both Ford and GM were buying Volvo Car and Saab Automobile to *learn to build quality cars*. It did not help. And Daimler Benz' attempt in the 1980s to build a gigantic "technology power house" turned out an equally gigantic flop.

Identifying a great *business opportunity* for both parties to a marriage? I have no successful examples from the automotive industry. For a while I was tempted to say that the marriage between a tiny sports car maker, such as Koenigsegg or Spyker with Saab Automobile would inject the marketing mentality that would allow technology smart Saab Automobile to raise its margins and stay small. However, that opportunity seems to have drowned in the financing mess and political overtures concerned only with the local employment consequences, and legal disputes over technology rights with GM. The needed financier with deep pockets and the right business competence does not seem to be around.

For a while the enormous platform investments needed to satisfy new environmental regulation and fuel efficiency in engine development were thought to become killing experiences for the small automakers. But those concerns appear to have been mistaken. Increasing oil prices have forced the industry onto a fuel efficiency mode, and electric cars are believed by some to define the future, but how, and to what extent, depends upon how the energy is to be stored in the car. Maybe that technology will be developed elsewhere and wreak havoc on the traditional automakers when small new upstart firms enter the market? Enormous resources have therefore been invested in building the car of the future even though the nature of the critical drive chain technology of that car is up in the air. The convergence to something is entirely experimental and the enormous new infrastructure that is needed for the economics of the new power supply that will have to be rebuilt entirely if the change is radical, won't be invested until the market knows with some certainty what the new system will look like.[27] This catch will not easily be resolved. The interesting question is whether you need enormous resources to solve that problem, or if the right idea will do it?

Creative destruction is typical of an Experimentally Organized Economy (EOE). We also know that *under the assumptions of the EOE there are always better allocations of resources than the current one*. Sweden currently appears to be on the verge of following the British example and letting the market decide how the technology assets left after the destruction will be picked up and recommercialized. The problem is that when large businesses with thousands of employees are involved the market adjustment mechanisms tend to be politically held back through regulation and subsidies. Old structures are conserved and new ideas locked up.

Looking back in history one can, however, see that a large number of automobile manufacturers have disappeared or been acquired by still surviving auto makers. In the 1960s GM alone had about 60 percent of the US automotive market which is now dominated by foreign producers. So what is best for the national economy? Is the disappearance of the domestic British automobile and transport equipment industry in the 1970s, notably helped along by Japanese price competition, an optimal solution? It has been "replaced" by an apparently viable specialized car and automotive engineering consulting industry, some of it being oriented towards racing car technology. Perhaps one should not worry so much about the fate of the Swedish automotive industry, but rather ask what the policy consequences will be if the large European volume producers PSA Peugeot Citroën, Renault, Fiat and Opel fail. They are all complaining about aggressive competition from superior South German automakers.

When a large number of giant players in excess supply markets approach their competition problems of multidimensional complexity with single-valued strategies, most of the time being the same strategy, but all different from time to time, the setting is that of a messy experimental selection process, in which some small players with deviant strategies may come out as winners. The total dynamics becomes one of immense complexity, and the policy concerns boil down to the question of where this experimental process will converge. Its national and regional impacts may be both devastating and constructive and the policy makers will have nothing to contribute except adding more to the mess by entering their special concerns. The risk of the automotive industry illustrated in this study and also expressed by John Reed (in FT 4 March 2008) "is car makers' tendency to pile into every new (conceived) opportunity en masse, leading to over capacity." However, he observes, automotive has been through more dramatic crises before and then demonstrated great robustness, so "global car production promises generous returns to companies that make the right (investment) choices." The question may even be whether the current fad to focus on small automobiles is the solution for the future (FT 5 May 2010) when upscale customers will continue to prefer large automobiles, and a new power supply technology may be waiting around the corner. The risk, however, is that the Swedish, like once the UK automotive industry, might vanish in the process, and that there is great uncertainty about what will come instead.

Notes

1 This paper is a direct sequel to the companion paper Eliasson 2011, and is based on interviews with and business journal clippings on firms documented in Eliasson 1996a and b, 2005a and b, 2006, 2010, Eliasson and Eliasson 2005a and b and other, not directly referenced, sources.

2 In order not to complicate the language, I use the term "volume" in a loose popular sense. I am of course aware that scale has to be related to the size of the market.

Volume, furthermore, as a statistical national accounting concept, includes quality supplied corrected for inflation if price indexes have been properly corrected for quality change, but it would mean very awkward language to keep that awareness explicit through the text.

3 Marshall in fact invented the notion of an industrial district to solve the dilemma of the Walrasian model of being incompatible with increasing returns, which was also Romer's (1986) problem, even though he did not quote Marshall.

4 In a national or global economic perspective the systemic productivity effects, or networking externalities, associated with distributed and integrated production that are forcing organizational change are not only based on the information, communication and coordination potential of C&C technologies, but also on the existence of functioning, high capacity transport networks that allow for stable, high speed, predictable and flexible flows of physical products, notably road transports.

5 After having been acquired by Daimler Benz in a failed merger attempt.

6 "The notion of a world car suitable for customers everywhere is the holy grail for car makers" writes the *Financial Times* (1 December 2010). Ford tried in the 1990s and failed. GM tried in the 1970s and failed. Now Ford is trying again on the presumption that "the World's consumers are becoming alike", and therefore all demand One Ford (*Time*, 6 September 2010). True, new C&C technology is making the one car strategy possible. The question is whether increasing numbers of upscale customers all want to drive the same car.

7 This venture in the early 1980s was, however, taken seriously. BMW felt compelled to acquire part of Messersmith- Bölkow-Blohm (MBB), Germany's largest defense contractor and loaded with aircraft technology, in order "to avoid being flattened by Mercedes", as Business Week expressed it then (9 December 1985). BMW, however, soon backed out, and MBB was acquired by DASA (Daimler Chrysler Aerospace) in 1989. In 2000 EADS was formed when DASA, Aerospatiale-Matra and CASA merged into the second largest aerospace company in the world.

8 From Volvo's history, as presented on the Web in March 2010.

9 Part of the hypothesis of the new version of new growth theory. See, for instance, Jones and Williams (1998).

10 The question is for how long. Until the Chinese have learned the Volvo know-how, or longer? Svenska Turbinfabriks AB Ljungström (STAL) was founded in 1913, acquired by ASEA (now ABB) in 1984, by French Alstom in 2000 and by German Siemens in 2003. On 22 August 2011 CEO Jan-Erik Ryden (DI, 22 August 2011: 15f.) announced that Siemens Industrial Turbomachinery in Finspång had experienced no problem with current market turmoil, had doubled its sales in the last five years, and expected to increase sales an additional 50 percent by 2015. One week later CEO Ryden learned (DI, 30 August 2011: 13), that troubled, fragmented and scandalized Siemens (FT, 23 April 2006: 21), headquartered in Munich, had decided to move production to Görlitz in Germany.

11 Volvo started this plant in 1989 with Government subsidies to offer jobs for laid-off shipyard workers. The plant was shut down in 1993, but restarted in 1995 together with TWR.

12 Volvo and Pininfarina split in 2011. The Uddevalla factory will be shut down in 2013, and whatever remains of production moved to China within the new Chinese owned Volvo Car (DN Ekonomi, 4 October 2011: 20f).

13 The company's ancestors are credited with making the first automobiles in the world: a three wheel car by Karl Benz in 1885, and a four wheel car in 1886 by Gottlieb Daimler.

14 In the wake of the cold war disarmament euphoria, tens of thousands of electronic and software engineers had to emigrate out of the downsized Californian military equipment industry, to revolutionize, notably, Hollywood animation studios and Detroit automotive design laboratories (Eliasson 2010: 159).

15 The BMW logo shows a stylized propeller.
16 The main reason for abandoning Rover was that BMW engineers had to spend excessive time fixing outdated Rover facilities and automotive models that took attention away from, and began to threaten, BMWs own product development.
17 In 1934 the firm received an order to design and construct a German Volkswagen. The first Beetle was built a year later in the garage of Ferdinand Porsche's home in Stuttgart. The first Porsche sports car was built in 1948.
18 Just as a footnote, Swedish super-car developer Koenigsegg also claims to have made a series-built car that tops 400.
19 Saab, including its automobile division, merged with Scania in 1969 to become Saab Scania. Both firms belonged to the Wallenberg group of firms. In 1996 the dominant owner Investor took Scania back to the Stockholm Stock exchange as an autonomous company.
20 The problem is that GM 2005 located all technical development of platforms and architectures used in several models in a separate company (Global Technology Development) from which all GM cars license the technologies, as do Saab. (*SvD Näringsliv*, 8 November 2011: 4f.; *Ny Teknik*, Nr 45, 9 November 2011: 6).
21 Founded in 1930. Volvo Flygmotor changed its name to Volvo Aero Corporation (VAC) in 1994.
22 See, for instance, the series of Ford history articles in *The Financial Times*, 29, 30, 31October 2007.
23 The Toyota way has its limits. Even though Toyota passed GM in size to become the world's largest automotive maker in 2010, it lost control of its overstretched supply chain, and experienced serious product safety design flaws. That blow to its quality image involved far more negative value for the company than the small benefits gained by becoming larger (*The Economist*, 27 February 2010).
24 Even though the top management of Volvo talks about a successful integration of Volvo Truck with Renault and MACK (DI 8 July 2010).
25 Once taximeter manufacturer Halda (now sold) was reorganized within Wallenberg controlled Incentive. It acquired special steel manufacturer Garphyttan (founded 1905), and has successfully developed four wheel drive systems for Volvo and the SUV platform for Ford.
26 Founded 1856, acquired by Volvo 1942. In 2004 Volvo formed *Gertrag All Wheel Drive* in Köping, together with Gertrag and US Dana, the latter two owning 60 percent (Eliasson and Eliasson 2005b: 453f.; DI 3 September 2011).
27 Volvo Car has decided to go for an electric car, but has already changed technology strategy twice In September it decided to leave Swiss Brusa and Canadian Magna to engage in a different project with Siemens to develop a plug-in hybrid engine for its C30 Electric (DI, 1 September 2011: 15).

Bibliography

Ackoff, R.L., 1967, "Management Misinformation Systems", *Management Science*, Vol. 14, No. 4 (Dec).

Antonov, Mikhail, and Georgi Trofimov, 1993. "Learning through Short-Run Macroeconomic Forecasts in a Micro-to-Macro Model". *Journal of Economic Behavior and Organization*, 21 (2), June.

Dearden, James, Barry W. Ickes and Larry Samuelson, 1990, "To Innovate or not to Innovate: Incentives and Innovation in Hierarchies", AER Volume 80, Number 5 (December): 1105–1124.

DI: *Dagens Industri*

DN: *Dagens Nyheter*

Eliasson, Gunnar, 1976. *Business Economic Planning, Theory, Practice and Comparison.* London, New York: Wiley & Sons.

Eliasson, Gunnar, 1996a. *Firm Objectives, Controls and Organization – the use of information and the transfer of knowledge within the firm.* Boston/Dordrecht/London: Kluwer Academic Publishers.

Eliasson, Gunnar, 1996b. "Spillovers, Integrated Production and the Theory of the Firm", *Journal of Evolutionary Economics*, 6, 125–140.

Eliasson, Gunnar, 2003. "Global Economic Integration and Regional Attractors of Competence", *Industry and Innovation*, 10, 75–102.

Eliasson, Gunnar, 2005a, "Competence Blocs in the Experimentally Organized Economy", in Eliasson, Gunnar, 2005, *The Birth, the Life and the Death of Firms –the role of entrepreneurship, creative destruction and conservative institutions in a growing and experimentally organized economy*, Stockholm:The Ratio Institute.

Eliasson, Gunnar, 2005b, "The Nature of Economic Change and Management in a New Knowledge Based Information Economy", *Information Economics and Policy*, 17: 428–456.

Eliasson, Gunnar, 2006, "From Employment to Entrepreneurship", *Journal of Industrial Relations,* 48(5): 633–656.

Eliasson, Gunnar, 2009, "Knowledge Directed Economic Selection and Growth", *Prometheus*, Vol. 27, No. 4 (December) 2009.

Eliasson, Gunnar, 2010, *Advanced Public Procurement as Industrial Policy – Aircraft industry as a technical university*, New York, Dordrecht, Heidelberg, London: Springer.

Eliasson, Gunnar, 2011, *The Internet as a Global Production Reorganizer.* Paper (revised) prepared for the 13th Conference of the International Joseph A. Schumpeter Society, Aalborg, 21–24 June 2010. To be published in Pyka and Andersen (eds), *Long Term Economic Development, Economic Complexity and Evolution*, Berlin and Heidelberg: Springer-Verlag.

Eliasson, Gunnar and Åsa Eliasson, 2005a. "The Theory of the Firm and the Markets for Strategic Acquisitions", in Cantner, U., Dinopoulos, E. and Lanzilotti, R.F. (eds), *Entrepreneurship: The New Economy and Public Policy* (Springer: Berlin, Heidelberg, New York).

Eliasson, Gunnar and Åsa Eliasson, 2005b "The Lake Mälar Competence Blocs – a Crisis and Opportunity Analysis of Regional Industrial Restructuring" Chapter VIII in Eliasson, Gunnar, 2005, *The Birth, the Life and the Death of Firms-the role of entrepreneurship, creative destruction and conservative institutions in a growing and experimentally organized economy,* Stockholm: The Ratio Institute.

FT: *The Financial Times.*

Hanusch, Horst and Andreas Pyka, 2007, "Principles of Neo-Schumpeterian Economics", *Cambridge Journal of Economics*, 31: 275–289.

IHT: International Herald Tribune.

Jones, Charles, I. and John C. Williams, 1998, "Measuring the Social Returns to R&D", *The Quarterly Journal of Economics,* Vol. 113. No. 4 (Nov): 1119–1135.

Jones, Charles and John C. Williams, 1999, "Too much of a good thing? The economics of investment in R&D", NBER Working Paper Nr. 7283 (August), Cambridge MA:NBER.

Klepper, Steven, 2006, *Strategic Disagreements, Spinoffs, and the Evolution of Detroit as the Capital of the US automobile Industry,* Department of Social and Decision Sciences, Carnegie Mellon University (February).

Knight, F., 1921. *Risk, Uncertainty and Profit*. Boston: Houghton-Mifflin.

Marshall, Alfred, 1890. *Principles of Economics*. London: Macmillan & Company.

Marshall, A., 1919. *Industry and Trade*. London: Macmillan & Company.

McKnight, Lee W. and Joseph P. Bailey (eds), 1997, *Internet Economics*. Cambridge, MA. and London, England: The MIT Press.

Okamuro, Hiroyuki, André van Stel and Ingrid Verheul, 2011, Explaining Differences in Entrepreneurial Activity between a managed and an Entrepreneurial Economy: Political Lessons from Japan and the Netherlands, International council for Small Business 2011 Conference in Stockholm: http://icsb2011.org/.

Patel, Parimal and Keith Pavitt, 1994. "The Continuing, Widespread (and neglected) Importance of Improvements in Mechanical Technologies". *Research Policy*, 23, 535–545.

Pratten, C., 1976. *A Comparison of the Performance of Swedish and UK Companies*. Cambridge: Cambridge University Press.

Romer, P.M., 1986, Increasing Returns and Long-Run Growth, *Journal of Political Economy*, Vol. 94, No. 5 (Oct.), 1002–1037.

Schumpeter, J.A., 1911, *Theorie der Wirtschaftlichen Entwicklung*, Dunker und Humblot, Jena. English ed., 1934, The Theory of Economic Development: An Inquiry into Profits, Capital, Credit, Interest and the Business Cycle, Vol. XLVI, Harvard University Press, Cambridge, MA.

Sv.D: *Svenska Dagbladet.*

Westerman, J., 1768, *Om Svenska Näringarnes Undervigt emot de Utländske, förmedelst en trögare Arbets-drift* (On the inferiority of the Swedish compared to foreign manufacturers because of a slower work organization), Stockholm.

8 Innovation policy, innovation and innovation finance co-evolution

The Israeli case and some implications

*Morris Teubal**

1 Objectives

This paper adopts *a broad co-evolutionary framework* linking the evolution of Innovation Policy on the one hand and Business Innovation and SU evolution and VC/Innovation Finance evolution on the other. The focus is on the Israeli experience, which which might illuminate the type of processes that may occur elsewhere. It combines two approaches of the evolutionary economics literature: that which searches for mutual links between aggregates such as the co-evolution between technology and institutions pioneered by Nelson (1994, 2001, 2007); and links among agents in the complexity literature, e.g., Allen's 2004 piece (Allen 2004). This paper will strongly suggest that co-evolutionary processes in Israel contributed to the emergence during 1993–2000 of a VC industry and market and of an entrepreneurial, ICT-oriented high-tech cluster (EHTC). They began at least two and half decades earlier, with the creation of a specific agency in charge of direct support of R&D in firms (OCS-Office of the Chief Scientist, Ministry of Industry and Trade, and its Grants to BERD program).

The specific objectives are:

- i Present a broad co-evolutionary framework of analysis linking Innovation, Innovation Policy, and Innovation Finance, focused on the Israeli experience during 1969–2000;
- ii Link the above with the emergence of a high impact VC industry and market and, related to this, the emergence of an EHTC between 1993–1996/7 (Phase 3 below);
- iii Suggest some possible implications of the analysis, in particular in relation to an Innovation and Structural-Change led process of Economic Growth.

2 Innovation policy in Israel: a three phase model

Our starting point is the evolution of innovation, high-tech and innovation policy in Israel during 1969–2000, the central features of which are summarized in Table 1.[1] The processes that took place during the emergence Phase (Phase 3) were much faster than those which took place in the previous eight years (Phase 2, see the middle column of Table 8.1) and much more so relative to the

Table 8.1 Israel's high-tech cluster: selected structural elements

A: Accumulated figures for a period	*1969–1984*	*1985–1992*	*1993–2000*
Number of high-tech start-up creation (VC Backed*)	136 (0)	349 (23)	2,436 (855)
Israeli VC Fundraised / VC Invested in Israeli start-ups (M$)	0/0	~85/~50	7,480/~5,600
Number of IPOs at US (at EU & TASE)	14 (7)	19 (15)	101 (75)
Number of significant Trade Sales (M&As)	0	2	91
Amount Raised: public markets & M&As (B$)	0.3	0.8	36.7
B: Figure for an actual year	*1984*	*1992*	*2000*
Share of ICT in Manufacturing Exports	14%	28%	53%
ICT Exports M$ (as % of ICT sales)	~900 (50%)	2,711 (50%)	12,893 (59%)
Software Development Exports M$ (as % of software sales)	5 (4%)	135 (23%)	2,600 (70%)
ICT professional employees (000)	~42.9	61.7	152.4
Patents Issued in the US (ICT Patents Issued)	193 (44)	355 (89)	969 (417)
R&D as % of GDP (OCS R&D Grants M$)	2.4% (97)	2.6% (199)	4.5% (440)

Sources: Avnimelech & Teubal, various papers; IVC (2008) – row 2-6; CBS (2008) – row 8-11 (estimates from IAEI); USPTO (2008) – row 12; and OCS (2008) row 13.

Notes

* By investment year.

Notice that in Table 8.1A, the second column represents 16 years, while columns three and four represent 8 years.

processes which took place during 1969–1986 (Phase 1).This is consistent with the view that the (accelerated) emergence of these new Higher Level Organizations was a Market-Dominated Endogenous process, one fuelled by, among other things, Innovation (and other) Policies, throughout the whole 1969–2000 period.

Box 1 outlines the conceptual *Three Phase Innovation Policy model* which underlies the above table. The three phases represent the Innovation Policy component of the corresponding three phases of VC's Industry Life Cycle till the year 2000 (Avnimelech and Teubal 2006). Thus, the first Innovation Policy phase took place during the VC's 'Background Conditions Phase'; the second policy phase during VC's 'pre-emergence phase'; and the third during the VC and Entrepreneurial High-Tech Cluster (EHTC) Emergence Phase (Avnimelech and Teubal 2006a). For each phase we can find (i) a summary of the Innovation Policy programs; (ii) the direction of their direct impacts; and (iii) some of the more 'dynamic' impacts particularly those favoring transitions to the subsequent phase.

Box 1: Phases in the emergence of a VC market and Entrepreneurial High-Tech Cluster (EHTC) in Israel: innovation policies and their impact

Phase I: Diffusion of R&D and Generation of Innovation Capabilities (1969–1984)*

- ffi*Creation of the OCS

Horizontal Grants to BERD program → Creation of R&D-performing companies, of R&D/Innovation capabilities, and creation of civilian High-Tech industry and first SU companies (first half of the 1980s)

Phase 2: Strengthening of Business Sector R&D and SU/VC Experiments (1985–92)

- 1984 R&D Law strengthens the OCS role in financing BERD
- Sharp Increase in Business Sector R&D grants. Also, Incubator and Magnet program (early 1990s, supporting cooperative, generic R&D)
- Restructuring of Defence industries including Defence R&D which also focused on civilian-relevant areas such as communications, etc.
- Business Sector R&D expansion → Increased rate of SU formation
- Business Experiments and Informal VC activity → New Model of SU ('born global' with links to global capital/product markets)
- While no private and professional VC market existed, a variety of SU-support mechanisms were used or were experimented with, including Angels, OCS subsidies, a few private VCs, tax concessions to company R&D, a special form of VC oriented to finance groups of projects rather than firms, etc.
- A failed VC support program (Inbal). By learning from Inbal's failure and from Business Experiments → Identification of System Failure (absence of

significant VC) and Selection of Limited Partnership (LP) form of VC Organization

- A critical mass of about 300 SUs became available by 1992, some of them of high quality (a few having IPOs NASDAQ) → increased Demand for VC services → Once VC funding was made available, it could trigger a market-driven, virtuous VC-SU co-evolutionary process

Exogenous factors and Global Context: Liberalization of global communications markets; new possibilities of immigration from the former Soviet Union, beginnings of the Software Industry, etc.

General Policies: Liberalization of Trade, Capital Markets, Foreign Exchange market, etc.

Phase 3: Targeting VC and an ICT-oriented, high-tech, Entrepreneurial Cluster; and Accelerated Growth of R&D and High-Tech (1993–2000)
Targeted Support of VC (Yozma Program); continuation of all Innovation Policy programs, R&D Grants peaked in 2000 → Emergence of a VC industry and Entrepreneurial Cluster → Accelerated growth of SU segment and High-Tech; large numbers of IPOs and M&A, etc.

* The names of the phases reflect the main objectives of Innovation Policy. *Grosso modo* there was consistent STE support throughout the Three Phases.

2.1 Phase 1 Background Conditions: Diffusion of R&D and Generating Innovation Capabilities (1969–1984)

The Horizontal Grants to Business Sector R&D program began in 1969 with the creation at the Ministry of Industry and Trade of a specialized agency, the Office of the Chief Scientist (OCS). This program was and continues to be the backbone of the country's R&D/Innovation strategy. Until the early 1990s, more than 90% of OCS disbursements to Civilian R&D came from this program, which supports the R&D activity of individual companies oriented to new/improved products and processes directed to the export market.[2] In contrast to a targeted program which is applicable to a specific industry or technology a Horizontal Program is open in principle to all firms whatever their sector; and to all R&D projects whatever their product class and technology. Horizontal programs of this kind are market friendly R&D support programs which give primacy to the bottom-up identification and generation of projects. In Israel it extended a 50% subsidy to every R&D project accepted by the OCS, regardless of the firms' industry, product class & technology (Teubal 1983, 1997).

The major objectives of the Horizontal R&D Grants Program during early implementation was to promote collective learning about R&D/Innovation,[3] to promote technological entrepreneurship, and to generate knowledge about potential areas where the country concerned might have or could develop a sustainable competitive advantage. R&D-performing firms mutually learn from each other, and a lot of this learning relates not directly to technology or R&D proper but to organizational and managerial factors. Box 2 provides a categorization of intra-firm learning processes, and instances of collective learning. Both are based on the Israeli experience for the 1969–1990 periods.

Box 2: Learning Process (Phase I)

Intra-firm Learning during Horizontal Program implementation-early sub-period
Learning How to search for Market and Technological Information.
Learning How to identify, screen, evaluate, choose and configure new projects
Learning How to generate new projects, including more complex ones
Learning How to manage the innovation process (linking Design to Production and Marketing; Selection of Personnel; Budgeting; Management of Human Resources etc.)

Collective learning
Firms learn about the importance of marketing
Firms learn how to establish and manage Strategic Alliances both with domestic and foreign companies, and how to generate links to Global Markets
The Office of Chief Scientist and the firms learned how to better assess the quality and economic potential of various types of projects and about R&D-related areas with potential Sustainable Competitive Advantage

2.2 Phase II Pre-emergence: Strengthening of Business Sector R&D and SU/VC Experiments (1985–1992)

The 1984 R&D Law further consolidated Israel's support of business sector R&D. The objective was to support knowledge intensive industries, which also required expansion of the science and technology infrastructure and exploitation of existing human resources. Another objective was creation of employment including absorption of immigrant scientists and engineers, etc. The outcome was a significant increase in Horizontal R&D awards to industry, and the emergence of software as an industry (see Breznitz 2007, 2008). Box 3 and Table 8.2 bring data on new policies initiated in Israel during Phases 2 (which policies continued during Phase 3). The table also brings data on the backbone Grants to BERD program, which was implemented throughout the three phases.

Since VC/cluster emergence in Israel was 'policy led' (Avnimelech and Teubal 2006) and since there were strong links between the earlier Grants to BERD program and the later Yozma program (Avnimelech and Teubal 2008a), it is important to understand the nature and context of the policies undertaken. New national priorities emerged with the arrival of the massive immigration from the former Soviet Union during the late 1980s–early 1990s. The government of Israel began searching for means to employ the thousands of engineers that came to this country. Simultaneously, the military industries had laid-off hundreds of engineers, and there were many attempts to create startups, which largely failed. One of the new objectives was enhancement of startup formation, survival, and growth.

Until the 1990s, the percentage of successful young companies was extremely low and the accepted view was that this resulted from weak management skills. Experts in the field realized that despite massive government support for R&D there still was a clear 'market failure', which blocked the successful creation and development of startups. The basic problem was lack of capability to grow after the product development phase. The head of OCS at the time, Yigal Erlich, thought that the missing link was marketing and management skills, and that the way to get it was to foster VC.

A new phase in the globalization process began during the late 1980s: foreign startups could more easily float in Nasdaq, provided the economy had adapted to the new opportunities of the ICT revolution. Part of Israel's adaptation involved new government programs, which complemented the 'regular' Grants to R&D (in firms) support program. These included *Targeted* programs supporting VC (Inbal and Yozma), and complementary programs raising the demand for VC services (e.g. Technological Incubators' Program, and the increasing regular R&D program which added tax exemptions to skill intensive companies). Simultaneously, foreigners and Israelis began searching for new investment opportunities in Israel; and three (pre-Yozma and pre-Inbal) VC companies began operating (Athena/Veritas, Star, and Evergreen).

Box 3: Phase 2: New Innovation and Technology Policy Programs

***BIRD-F (1981–)** a quantitatively small program supporting joint innovation efforts between an Israeli firm (mostly performing the R&D part) and a US firm (mostly performing the production and marketing part). This program was widely emulated by other countries

*Inbal (1991) – a Government-owned insurance company, which gave partial (70%) guarantees to traded VC funds. Four VC companies were established under Inbal regulations. This early VC support program failed to create a VC industry or market

***Magnet Program (1992–)** – a $60M a year Horizontal Program supporting cooperative, generic R&D involving two or more firms and at least one university.

***Technological Incubators (1992–)** – a 25–30M$ program supporting entrepreneurs during the Seed Phase, for a period of 2 years. The incubators were privately owned and managed. Both they and the projects get financial support from the Government.

Table 8.2 OCS grants (million dollars) 1991–2003

Year	*Total Grants (Growth)*	*Grants to BERD (Individual firms)*	*MAGNET Budget*	*Technology Incubators*	*Royalties* (Growth)*	*BIRD-F Awards*
1985	106 (2.5%)	106	0	0	6 (33.3%)	NA
1986	110 (2.8%)	109	0	0	7 (16.7%)	NA
1987	113 (2.7%)	112	0	0	8 (14.3%)	NA
1988	120 (6.2%)	118	0	0	9 (12.5%)	NA
1989	125 (4.2%)	122	0	0	10 (11.1%)	NA
1990	136 (8.8%)	133	0	0	14 (40.0%)	NA
1991	179 (31.6%)	171	0	4	20 (42.9%)	12
1992	199 (11.2%)	177	1	16	25 (25.0%)	10
1993	231 (16.1%)	199	40	24	33 (32.0%)	12
1994	317 (32.2%)	172	10	27	42 (27.3%)	10
1995	346 (9.1%)	294	16	31	56 (33.3%)	12
1996	351 (1.4%)	279	36	30	79 (41.1%)	13
1997	397 (13.1%)	309	53	30	103 (30.4%)	12
1998	400 (0.8%)	305	61	30	117 (13.6%)	14
1999	428 (7.0%)	331	59	30	139 (18.8%)	9
2000	440 (2.8%)	337	67	32	135 (10.8%)	8
2001	431 (–2.0%)	328	64	32	145 (5.2%)	11
2002	383 (–11%)	291	58	27	153 (1.4%)	10
2003	369 (–3.4%)	283	53	26	133 (-5.4%)	11

Source: Office of the Chief Scientist. From: Avnimelech (2004).

Note
* From sales of products resulting from Grants to BERD projects financed by the OCS.

2.3 The Emergence Phase: the Yozma Program

New National Priorities emerged in Israel with the beginnings of the massive immigration from the former Soviet Union during the early 1990s. The Government began searching for means to employ the thousands of engineers that came to this country. Simultaneously, the Military Industries had laid-off hundreds of engineers; and many SU companies were created only to subsequently fail. In fact, an official report (a Jerusalem Institute of Management report of 1987) mentions that 60% of the technologically successful OCS-approved projects failed to raise additional capital for marketing and had to close the business.[4]

Officials in the Treasury and the OCS concluded that despite massive Government support for R&D there were clear 'market and system failures', which blocked the successful creation and development of Startup companies. As a result, a shift in policy objectives gradually took place – from promotion of R&D to enhancement of SU formation, survival and growth. System failures related not only to insufficient sources of R&D follow-up finance but also to weak management abilities, business know-how and non market-directed developments. Eventually policy makers believed that the way to overcome these deficiencies was to foster a domestic Venture Capital industry which then became a Strategic Priority of the Government of Israel.[5]

The first VC targeted program was Inbal (a failed program supporting *Public VC funds,* raising capital at the Tel Aviv Stock Exchange, TASE), whose implementation started in 1992. The second one was Yozma, a successful program implemented during 1993–1997. As mentioned, this program was credited with triggering the creation of a domestic VC industry and market. Tables 8.3 and 8.4 show the strong acceleration of VC, and ICT activity during the 1990s, while Table 8.5 shows that of start-ups.[6]

Table 8.3 Venture capital raised and invested and SU foundations

Year	*VC Raised*	*SU Foundations (VC Backed)*	*VC Invested (% of foreign)*	*VC investment as % of GDP*
1991	40	53(4)	NA	NA
1992	49	94(12)	NA	NA
1993	82	124(73)	NA	NA
1994	112	140(85)	NA	NA
1995	135	175(87)	NA	NA
1996	317	231(117)	293(32%)	NA
1997	637	260(119)	436(40%)	0.41%
1998	653	312(152)	589 (43%	0.54%
1999	1,160	573(208)	1,011 (57%)	0.9%
2000	2,521	642(372)	3,233 (59%)	2.6%

Source: From Avnimelech & Teubal 2009 and IVA and Money Tree.

Table 8.4 Capital raised by PE/VC organization in Israel*

	1990	*1991*	*1992*	*1993*	*1994*	*1995*	*1996*	*1997*	*1998*	*1999*	*2000*
Private VCs	40	49	27	33	82	93	287	595	653	1,160	2,712
Yozma VC	0	0	0	149	40	15	30	19*	0	0	0
Inbal VC	0	0	54	22	0	0	0	0	0	0	0
Other PE	5	9	79	168	262	25	620	134	33	258	66
Total	45	58	160	372	384	133	937	777	686	1,418	2,778

Note

* In millions of US dollars. For details see Avnimelech & Teubal 2006.

2.3.1 The emergence process[7]

The rapid quantitative growth of startup and VC activity that took place during the *Emergence* process led to a well defined VC industry and market and startup-intensive, entrepreneurial high-tech cluster. Emergence is not a simple replication of selected agent types, since what emerges are *higher level organizational forms.* In our context, it is the outcome of an evolutionary process leading to *Structural change or to Qualitative Transformation & Structural Change* (Foster and Metcalfe 2001, Fig. 1.1, p. 6). Alternatively it could be stated that while post variation 'selection/development' introduced new quality into the system (e.g. the VC Limited Partnership and a well defined SU business model) its full economic impact required *scaling up* of the new entities.

In terms of complexity theory, Allen states that:

> in a system in which we do not make the reductionist assumptions of average types and average process rates … that takes out the natural micro-diversity and idiosyncracy of real-life agents, actors and objects, then we *automatically obtain the emergence of* structural attractors.… These are complex systems of interdependent behaviour whose attributes are on the whole synergetic.
>
> (Metcalfe and Foster 2004, Chapter 5)

This statement would be true if it would allow for the possibility of Left Hand Trunkation (Avnimelech and Teubal 2006) – that is, pre-emergence phase processes (e.g, variation and selection) which would not actually lead to emergence of the new, higher forms of organization.[8] Related to this it is important to emphasize that emergence processes are cumulative with positive feedback, i.e., dynamic economies of scale; and that 'small historical events' may lead to changes in the strength and pattern of emergence (or, as mentioned, lead to non emergence) i.e., they are *path dependent dynamic trajectories.*

The existence of dynamic increasing returns in the creation of *entrepreneurship systems* has been emphasized by Lerner in his book on policies promoting

Table 8.5 ICT and software manufacturing: sales, exports (thousands US$) and employees (1990–2002)

Year	*ICT Sales*	*ICT Exports*	*ICT Employees*	*Sale per Employee*	*Software Sales*	*Software Export*	*Software Employees*	*Sale per Employee*
1990	3,300	2,100	32,000	103	400	75	5,000	80
1991	3,600	2,280	33,000	109	540	110	5,000	108
1992	4,000	2,660	34,200	117	600	135	5,500	109
1993	4,600	3,200	36,400	126	700	175	6,200	113
1994	5,200	3,750	37,600	138	800	220	7,000	114
1995	5,900	4,300	39,200	151	950	300	7,700	123
1996	6,500	4,880	42,000	155	1,300	600	8,500	153
1997	7,200	5,700	43,700	165	1,780	1,000	10,000	178
1998	8,000	6,550	45,600	175	2,350	1,500	11,500	204
1999	8,600	7,130	48,000	179	2,950	2,000	13,000	227
2000	12,500	11,000	54,800	228	3,700	2,600	14,500	255
2001	11,250	9,750	47,000	239	4,100	3,000	15,000	273
2002	10,000	8,800	43,200	231	2,800	1,900	13,200	212

venture capital and entrepreneurship more generally speaking (Lerner 2009). The factors he mentions are shown (with minor additions in Box 4 below):

Box 4 Virtuous cycles in entrepreneurship and venture capital

Employees of large firms may initially[9] *be reluctant to join or create a SU*

Much of the entrepreneurial process is an art rather than a science. This would imply a learning by doing, and even a collective learning by *Others Doing process.*

Entrepreneurs learn about the tradeoffs associated with involving seasoned venture capitalists in their ventures (balance between terms and conditions for the investment and an appreciation of the type of gains possible).

Lawyers and accountants become familiar with the venture process and can better advise entrepreneurs and financiers alike.

Institutional investors gain confidence

Venture capitalists can more readily find peers with whom they can share transactions (syndication is very important).

In what follows we visualize what would have been the process during the various *sub-phases of Phase 3*. During the *fluid* sub-phase a continuation of the experimentation and collective learning of pre-emergence took place leading to further post selection/development specification of SU and VC management, strategies and organization.[10] It ended with a specific *technological focus* and with a specific set of business models and exit mechanisms of the companies in the cluster. During the *rapid growth* sub-phase, accelerated entry of startups and VC companies occurred, and a domestic VC industry was created. It is then that the cluster attained a size which enabled it a better exploitation of economies of scale in the domestic generation of non-traded (or partially non-traded) inputs to high-tech, e.g., accountants, lawyers, investment bankers, consultants, providers of knowledge inputs, and suppliers of production and marketing inputs. This factor, together with the existence and expansion of networks of personal and professional links, contributed to reductions in *transactions costs* and facilitated the access of agents in the international arena with critical marketing, reputation and other services. Through time Israel's ICT cluster became increasingly capable of providing effective services to new startups, both through VC and through other input and service suppliers. It also induced entry of multinational companies in the ICT areas and of additional domestic and (eventually) of foreign VCs, followed by rapid creation of new startups and rapid growth in the acquisition activities of multinationals. As long as external and internal conditions remain unchanged, the dynamic increasing returns to scale or positive feedback cumulative process of creation of large numbers of startups continued and the startup-intensive ICT cluster continued to grow. However, as the global ICT market entered the bubble-and-bust period, the local Israeli ICT cluster followed it. At this point the accumulated strength and stability of the cluster was crucial for future survival and development.

The trigger of VC emergence was implementation of Yozma in 1993 which led to $250M capital raised and significant amounts invested up to 1996 (Avnimelech and Teubal 2006, 2008b). This and a few successful exits during the mid 1990s induced further entry by other private VCs and through these, further SU entry as well. Moreover, the strong collective learning processes further stimulated entry both of new startups and of new VC companies, and accelerated creation of second and third funds both by existing Yozma VC companies and by other VC companies.[11] Finally, emergence was associated with strong VC-SU Co-evolution which by integrating other sub-processes (through indirect effects) became key to the overall process(see Section 5 below).

2.3.2 Policy targeting of venture capital: the Yozma Program (1993–1996/1997)[12]

There is wide consensus that one of the major factors triggering emergence of Israeli VCs was Yozma. Four sets of factors seem to have been responsible for *Yozma to become an effective trigger of Israel's new ICT Cluster:* (a) favorable background (Phase 1) conditions; (b) a critical mass of SU by 1993 which assured the existence then of 'investment ready opportunities' (Mason and Harrison 2003) i.e., a clear demand for the services provided by VCs;[13] (c) the high quality of some of these SU, e.g., Checkpoint, Ornet, Galileo, and M-systems who made a significant direct and indirect contribution to emergence; (d) policy and market forces' experimentation during the pre-emergence period; and (e) Yozma's timing and its successful design and implementation.

Let us recall additional background and pre-emergence conditions which *transformed Yozma into an effective trigger of Israel's new ICT cluster*: the 'regular' R&D support program which directly and through co-evolutionary processes led to the diffusion of innovation and to the creation of SU and other R&D-performing companies: the restructuring of the military industries, the massive immigration from the former Soviet Union during the 1990s, new global innovation opportunities opened by the ICT revolution and a *cultural shift* where entrepreneurship was increasingly being considered prestigious in Israeli society. All of these generated a spurt of startup activity during the early 1990s.

The Yozma program began operating in 1992 and the first fund was created in 1993. The explicit objective was to create a solid base for a competitive VC industry with critical mass of capital and activity, to learn from foreign limited partners, and to acquire a network of international contacts. It was based on a $100M Government owned VC fund (with the same name) oriented to two functions: (a) investment in ten private VC funds ($8M in each one) and (b) direct investment in high-tech companies – $20M. The basic thrust was to promote the establishment of domestic industry populated by Limited Partnership VCs that invested in very young Israeli high-tech startups with the support of government and with the involvement of reputable foreign VC investors. Each 'Yozma Fund' had to enlist one such foreign institution together with a well-established Israeli financial institution. The $100M of Government funds draw $150M of private sector funds (domestic and foreign) making a total of $250M which were invested in about 200 startups. Each Yozma fund had a call option on Government shares, at cost (plus interest) and for a period of five years. Thus Yozma did not simply provide supply, risk sharing incentives to investors, but it also provided an upside incentive (when the VC investments where profitable the private investors could leverage them through acquisition of the government shares).[14] The program also assured the realization of learning through the compulsory participation of foreign financial institutions. Box 5 below reproduces some of its main features.[15]

Box 5: Design of Yozma Program

- Fund of Funds (80% of Government VC contribution) and Direct investments in SU (20%).
- Favoured a Limited Partnership type of VC company.
- A focus on Early Phase investments in Israeli high-tech Startup companies.
- Target Level of Capital Aimed at 250M$ (Government Support- 100M$) – this was the 'Critical Mass' of VC supply required for VC industry 'emergence'.
- Ten Privately owned Israeli LP VC Funds each managed by a local management company (formal institution) and involving Reputable Foreign Financial Institution.
- Government Participation in each Fund – eight million dollars (up to 40% of fund's capital).
- Strong Incentive to the "Upside"- a five-year option to buy the Government's share (almost) at cost.
- The 'Privatization' of Yozma Fund and Program was completed in 1998 (Yozma became a Catalytic Program).
- Contributed to attract or induce the creation of a wide variety of agents such as MNEs, foreign investment banks, a variety of service providers, and eventually top tier foreign VCs.
- As part of emergence, we observe a strong process of VC-SU co-evolution during the 1993–67/8 period*.

*see section 5

3 Two co-evolutionary processes

The co-evolutionary processes to be analyzed in the various phases involve three variables: Innovation, Innovation Policy and Innovation Finance. *Innovation* is defined broadly to include also capabilities and innovative Business Sector organizations including SUs *(Innovation*)*; and *Innovation Policy* includes direct Government support of innovation in firms including joint STE/University-Business programs such as Grants to BERD, tax support of innovative firms (not considered here) and other policies such as support of VC or other private sources of innovation finance *(Innovation Policy *). Innovation Finance* includes private sources of finance of innovation and innovative firms such as VCs, angels and private equity companies excluding in-house funding by firms.[16]

Two co-evolutionary processes are considered:

(a): Innovation Policy*←-> Innovation*

and

(b): Innovation* and Innovation Policy←>Innovation Finance[17]

Innovation Policy-Innovation* co-evolution* requires that Innovation Policy, as in Israel, succeed in increasing business innovation, in developing innovation capabilities and in creating new R&D-performing firms, including SU (thus by 1974 there were more than 300 R&D-performing firms, for a program that started only a few years before with small budgets; while in 1993 there were approximately 300 ICT-related SUs). It also requires that the link between the variables be two-way. Thus in Israel, the expansion of capabilities and the enhanced awareness and willingness of entrepreneurs to harness R&D for company growth led both to modest yet continued increases in the budget of the OCS's Grants to BERD program and to the addition of new programs (e.g., BIRD-F).

The search for and adoption of private *Innovation Finance* mechanisms by the OCS and by other Government Agencies like the Ministry of Finance (MOF) started in the early 1970s and continued throughout the whole period analyzed in this chapter. It was the outcome both of increased 'demand' for innovation support by the business sector and of stringent Government budgets. By the late 1980s and early 1990s, and in response to large numbers of SU, the search and implementation of new, private Innovation Finance mechanisms led to Venture Capital becoming the central mechanism of innovation finance in Israel, and to an enormous increase in BERD/GDP ratio to over 4.5% during the second half of the last decade. Needless to say (b) is a two-way link-adoption of a new, private mechanisms of financing innovation stimulated innovation, innovative firms and eventually (especially during the 1990s) SU creation while also affecting – sometimes in conjunction with other factors – Innovation Policy, e.g., 'privatization of Yozma during the late 1990s, a relative reduction of the Grants to BERD program (private VC became the main funder of BERD in the economy, i.e., there was a clear substitution between the Grants to BERD program and VC funding of innovative companies and SU), and the enactment in 2005 of a program supporting innovation in traditional industry.

Both co-evolutionary links were present in Phases 1–3 of the evolution of Israel's VC market and entrepreneurial high-tech cluster. In what follows we illustrate some of these in more detail.

4 Co-evolution in phases 1 and 2

4.1 Co-evolution in phase 1

In this phase, Israel's entrepreneurial high-tech cluster did not yet exist. The Science, Technology and Higher Education infrastructure was strong and, stimulated by the OCS's Grants to BERD scheme which started in 1969, a segment of R&D-performing firms and a specialized labor force were being created. Throughout, OCS's proactive policy promoted 'demand' for R&D by firms both through a 'collective learning' process and through enhanced awareness of the commercial potential of R&D (Teubal 1997). Also, and in response to experience and changed circumstances, the OCS adjusted the Grants to BERD program and was instrumental in creating one or two new 'incentives' program (e.g., BIRD-F, see below). These processes led to the development and diffusion of R&D/innovation capabilities throughout the business sector, the beginning of global product and capital market links, creation of a favorable environment for foreign investment, the beginning of involvement of business sector agents such as banks in high-tech activity financing, and the very first steps of technological entrepreneurship activity. Also there were the first indications of regional high-tech industrial concentration(in Haifa).

4.1.1 Phase 1: innovation policy-innovation* co-evolution*

Two specific Innovation Policy programs and their impact illustrate Phase 1 *Innovation Policy*-Innovation* co-evolution*: creation of the Program Supporting *'Projects of National Importance'* and creation of the BIRD-F program. The former program was a response to the new innovation capabilities and opportunities of leading electronic firms of the 1970s, particularly Elscint, the outcome of the early years of implementation of OCS's Grants to BERD program. While financed by the World Bank, the new program was implemented through the OCS during the second half of the 1970s. It supported the R&D of 'large projects' involving not only a firm but also a University at an 80% R&D subsidy rate rather than the normal 50% of the Grants to BERD program (one of these supported Elscint's development of the CT Scanner, a highly complex project and product).[18] By promoting a higher degree of sophistication and complexity of R&D/Innovation projects and therefore facilitating future company growth and diversification, this program benefited those companies who had already accumulated substantial innovation capabilities. Through its success and by considerably enhancing 'demand' for R&D by firms it also indirectly spurred not only the growth of the OCS budget but also search for alternative sources of support for R&D/Innovation during the 1980s (e.g., two short-lived schemes namely the Elscint Law favoring Angel investment in groups of company projects, and a temporary scheme of tax benefits to firms based on company R&D expenditures).

Another new program was BIRD-F (Binational Industrial R&D Foundation) jointly promoted by the US and Israeli Governments with implementation starting in the early 1980s. BIRD-F supported joint *Innovation (rather than exclusively BERD) Projects* with the Israeli company focusing on R&D and the US company on accessing the complementary assets required for effective commercialization of the invention. As with the first program, BIRD-F depended on the innovation capabilities developed during the 1970s as a result of the OCS's grants to R&D program. It had an enormous influence in developing personal and business links between Israeli and US companies. Both programs directly and indirectly influenced innovation capabilities in firms and seemingly made an important contribution to the emergence of Israel's entrepreneurial high-tech cluster of the 1990s.

4.1.2 Phase 1:innovation (and innovation policy*)-innovation finance* co-evolution*

Specific links of this kind began even earlier than the program supporting 'Projects of National Importance'. The background was the strong rise in 'demand' for Government subsidies to R&D/Innovation derived from initial successes and collective learning of R&D-performing companies. A couple of early 1970s examples also involved Elscint. This company, who was considered Israel's flagship civilian high-tech company during the 1970s was an avid 'user' of OCS grants. Partly as a result of this and the excellence of its designs and products, a strategic agreement with General Electric involving the supply of Nuclear Medicine Scanners was made. Elscint's enhanced reputation generated from this agreement led to two other business agreements which increased the company's access to financial resources, over and beyond Government grants from the OCS: a loan from Israel Discount Bank (the only bank at the time to have made a strategic move to invest in the equity of high-tech firms) to Elscint, and the company's (and Israel's first) IPO in NASDAQ, in 1972.[19] Later on (mid/late 1970s) what was Elscint's problem increasingly became a problem of the whole R&D-performing business sector of Israel, namely that the limited funds available from the OCS to promote innovation were constraining the rapid increase in R&D projects/innovations desired and proposed by firms.[20] This led to a continued search for alternative mechanism of 'finance of R&D/innovation' during the subsequent decade of the 1980s.

Another important Innovation (and Innovation Policy)-Innovation Finance co-evolutionary effect of the early 1980s coincided with the beginning (world-wide and in Israel) of a separately identifiable Software industry during the early 1980s. The ample opportunities for new high-tech start ups (SU) that this entailed (partly also due to the significant R&D and Innovation capabilities accumulated till then as a result of OCS support) spurred the appearance of a new set of entrepreneurs and a new set of SU companies. These, which included, e.g., the Rad Group, Comverse (or its precursor Efrat), Amdox and others, were not linked to the Elron group which (largely)owned Elscint and other companies that emerged in the 1970s. The OCS was active, no less than directly financing these SUs, in finding alternative mechanisms of support as well. This led to the first and highly idiosyncratic form of Venture Capital financing in Israel, an important precursor to the high impact VC industry 15 years later. In this instance we meant by VC 'pools of money' fed from investments by private individuals in the US that were oriented to acquire stakes in *specific projects* or groups of projects (rather than to purchasing an equity stake in a company).[21] Israel's governmental policy was instrumental in implementing the new financial opportunities by helping investors, through loans, to take advantage of potential income tax breaks. The outcome was the beginning of a pool of domestic SU companies during the first half of the 1980s.

4.2 Co-evolution in phase 2

4.2.1 Phase 2: innovation policy-innovation* co-evolution*

Innovation policy disbursements continued to grow especially after the *R&D Law of 1984* which also provided a measure of autonomy of the OCS vis-à-vis the Ministry of Industry of Trade (thus generating more direct and effective links with the Ministry of Finance, in some cases leading to an effective 'weak constraint' to the OCS budget). This factor together with other policies – e.g., the *restructuring of the Defense industries starting in the mid 1980s* (a fact leading both to a significant transfer of Scientists, Engineers and Technicians and to an increased flow of new STE graduates to civilian industry) – also led, directly and indirectly, to the creation of new ICT-oriented SUs. The process was further reinforced with the enhanced accessibility of Israeli SU companies to NASDAQ starting in the early 1990s and other factors (see below), the effect being *that the numbers of SU operating in 1992 approximated 300*. We will see that the processes leading to growth of SUs had a strong effect on helping Israeli policy makers in the late eighties and early nineties to focus on the importance of creating a domestic VC industry, a critical step in the ***(b 2,3)*** co-evolutionary processes eventually leading to Yozma and to the emergence of a VC market/EHTC.

In parallel to the growth in SU numbers, we observe important learning effects and qualitative changes concerning the organization and strategy of SUs (particularly those involved in software and communications-related innovations) during the late 1980s and early 1990s. Continued experimentation and learning about R&D and innovation by SUs which benefited from the Grants to BERD or other programs (including those SUs founded by spinoffs from Defense) generated variety and eventual selection of adequate business models and organizational forms that fit the local context. This led to the *appearance and 'selection of' a new startup business model* which included a greater disposition to accept external investments which diluted owners' capital, a 'born global' outlook, and a strategy/organization which increasingly became oriented not only to product markets but also to knowledge and capital markets (both private and public).

The above SU-related developments and associated Innovation Policy-Innovation co-evolutionary links were underpinned by a new strategic priority of the OCS: promoting the growth and development of SUs. This, and the subsequent identification of system weaknesses/failures with respect to SU-support structures, directly induced policy makers to focus their attention on VC (late phase 2, early phase 3)

4.2.2 Phase 2: innovation-innovation finance* co-evolution*

The growth of R&D-performing companies which resulted from successful (and in large part) OCS-stimulated innovation, and the appearance of what could be termed a 'civilian' high-tech sector comprising companies mostly oriented to the US and European markets was also accompanied by the search of new mechanisms for financing innovation. An important 'new' source (beyond Elscint's pioneering use of this mechanism during the early 1970s) for the 'mature', non-SU companies that operated in the mid-eighties (Scitex, Tadiran, Optrotec/Orbo, Ormat, Teva and others) was undertaking IPOs in Nasdaq. New links with US underwriters and investment banks were formed, with the latter beginning to appreciate the innovative and associated capital market potential of Israeli companies.

These early links were two-way links, in the sense that the new financial possibilities further reinforced innovation capabilities and the growth of companies, but also the future possibilities of Israeli companies (both SU and more mature companies) to go public through NASDAQ. Lehman Brothers who was active as an underwriter of 'mature' companies in the mid 1980s became an underwriter for some of the IPOs of SUs during the early 1990s. This provided new opportunities for SU companies and must have been a factor explaining the acceleration of SU foundations in the early 1990s.

The above effects further contributed to create favorable conditions for the strong 'quantitative' SU-VC co-evolutionary process of Phase 3. To understand this we should go beyond the quantitative and qualitative effects on SUs surveyed above and consider VC as well. The growing importance of SU towards the late 1980s drew the attention of foreign investors/angels (e.g., Jay Morrison of California, who was co-founder of a VC company in Israel who later metamorphosed into Jerusalem Venture Partners) and foreign financial institutions who came to Israel (Advent, probably the leading Private Equity company at the time) to search for SU investment opportunities. To this we must add the creation of the first independent, Israeli ICT-oriented VC fund in 1985 (Athena) and two others later on. Thus, 'demand' for financial (and 'added value') services-from SU- generated 'supply', and vice versa.

At least for the software & communications areas where Israel excelled, a critical factor facilitating the above process was the *'mutual* qualitative *adaptation of the strategy/organization of the finance demand agents and the finance supply agents'* during the pre-emergence period (Phase 2). The qualitative changes in SU organization and strategy mentioned in 4.2.1 above, took place in sync with experimentation and mutual adaptation of various forms of 'finance supply' and more specifically, VC. The process could be described as leading to a clear preference for a VC-based financial intermediation 'model' involving independent, domestic organizations with a clear SU 'early phase' SU investment orientation. This led to selection of one particular VC organizational form-Limited Partnerships (the dominant VC form in the US and in Israel).[22]

Phase 2 summary and implications

In Israel's innovation and high-tech evolutionary history, a major condition for a successful transition to a *Pre-Emergence* phase (Phase 2) involving significant diffusion of R&D and innovation capabilities throughout the business sector, was the continued growth and adaptation of that country's Innovation policy portfolio and institutions, particularly but not only the OCS. Like in Phase 1, this caused *Innovation Policy** to co-evolve both with *Innovation** and with *Innovation Finance*;* and an ongoing *technological revolution* made the pool of technological opportunities upon which these processes depended to be continuously renewable.

A major component of *Innovation (and Innovation Policy) <–> Innovation Finance co-evolution was* an early form of *SU – VC co-evolution* which was *qualitative*, i.e., leading to 'selected' organizational forms & strategies of both variables. These qualitative changes (and other factors) contributed to the rapid growth of SUs during the late 1980s and early 1990s. The 300 SUs operating in 1992 represented a *critical mass* for the successful targeting of VC in 1993, whose impact was emergence of a domestic VC industry/market and entrepreneurial high-tech cluster.

A number of exogenous factors contributed to the above. Among those which were global we could mention the continued ICT-oriented technological revolution, globalization of NASDAQ and de-regulation of Communications and Communication input markets in major industrialized economies (which enhanced demand for communications-related hardware and software), creation of the semiconductor fabless model, and creation of the software and PC industries in the early 1980s. In parallel to these we should mention 'exogenous' domestic developments such as cancellation of the Lavi fighter plane project in the mid 1980s, and liberalization of foreign trade and capital markets.

5 Co-evolution during phase 3 and link with VC market and EHTC emergence

5.1 Phase 3 innovation policy-innovation co-evolution*

The expansion of the 'direct' Grants to BERD program and of the more recent Technological Incubator and Magnet Programs of the early 1990s, further reinforced SU creation and established R&D performing, ICT companies. These positive impacts and other factors, like the immigration of large numbers of engineers and medical doctors from the former Soviet Union and the 'taking off of the Internet' which led to the foundation of new firms like CheckPoint, led to further expansion of and 'diversification' within these programs[23].

5.2 'Quantitative' VC-SU coevolution

We mentioned that by 1992 there were about 300 SUs which represented a 'critical mass' for the eventual onset of a systematic SU-VC co-evolutionary process. Some of these were very good and successfully undertook an IPO in NASDAQ prior to 1993 (Lanoptics, Magic and The Third Dimension later on). The size of the overall 'stock' of SUs and the small share of those that were VC-backed at the time suggests the existence of *'unsatisfied startup demand for VC services'*.[24]

Over and beyond the expansion of existing programs which provided monetary incentives and financial services, these facts induced policy makers to search for alternative, private sources of finance. Two *private VC*-directed programs were implemented: Inbal in 1992, a failure; and Yozma in 1993, a big success. Both targeted private, independent Venture Capital funds, which were supposed to represent the '*supply side counterpart' to the above mentioned excess demand.* Once implemented, Yozma sparked a cumulative process of emergence where *Quantitative SU-VC co-evolution played a key role.* The outcome *was* emergence of Israel's entrepreneurial high tech cluster (EHTC) of the 1990s.

We may now summarize as follows. In contrast to the qualitative and rather sporadic SU-VC co-evolutionary processes of Phase 2, *Yozma's implementation starting in 1993 signals the onset of a more intense and systemic,* and largely *'quantitative' process of VC-SU co-evolution.* The direct and indirect links among variables (the latter through the intermediation of other components or variables of the emergence process) become increasingly regular and frequent.[25] *First* we observe a quantum jump in VC capital and a *'temporary excess supply of VC services'*, followed by a relatively rapid entrepreneurial response. As a result we observe not only an increased share of startups that were VC-backed but also significant increases in gross additions to startups during 1995–1996. These were either a reaction to the 'excess supply of VC services' or the expectation that new startup foundations would easily find *new* VC sources of finance if required. *Then* we observe *increasingly synchronous growth*: starting in 1996, startup demand for VC services and VC 'supply' became increasingly synchronous i.e. rapid mutual adjustment. During the rest of the decade, the share of startups that are VC-backed increased (up to about 50%).

6 Summary, conclusions and policy implications

6.1 Summary

In Avnimelech and Teubal (2009) it is concluded that:

> Introducing co-evolutionary processes in a much more extensive way than in previous papers of the authors considerably adds to the analysis of emergence (and to the evolutionary analysis in general) of Israel's VC industry and market and entrepreneurial ICT cluster. Despite numerous pieces of research on the topic of co-evolution and considerable advances made (see Nelson 1994, 2007; Murmann 2002, 2003; Breznitz 2007, 2008; Avnimelech and Teubal 2005; among others) the 'Schumpeterian' literature on co-evolution including this paper lacks a basic analytical framework for incorporating such processes into a broader evolutionary process. This paper

> suggests that an early, rather random and haphazard, largely qualitative co-evolutionary process among broad sets of variables ('R&D-performing companies' and 'Mechanisms of innovation finance') should precede a subsequent, more focused and synchronized process among a sharper and more reduced set of variables (startups and VC). The former *qualitative* co-evolution corresponds to our *pre-emergence phase* while the latter rather *quantitative* and *systemic* co-evolution corresponds to the *emergence phase*. Systemic VC-startup co-evolution would be the backbone of the rather complex process of emergence of the new, higher order organizations namely the VC market/industry, and startup-intensive ICT cluster.

Box 6 summarizes these and other aspects of the overall process from a slightly broader perspective.

Box 6: Types of Co-evolutionary (*CE*) Processes: Israel 1969–2000. (Inn=Innovation; Var=Variation; Sel=Selection; ^ means 'selected')

Phase	*Type of co-evolutionary(ce) process*	*Process characteristics*	*Impact of innovation policy, ce & other factors*
1 & 2	Strong & frequent Inn Policy*<→Inn* co-evolution Beginning of Inn*(& Inn Policy*) – Inn Finance* co-evolution	'Qualitative' Mutual Adaptation of Inn (esp. SU) – Innovation Finance* (esp. VC strategy/ organization) i.e. Simultaneous Var and Sel of SU & VC types ®{SU^ & VC^}	Diffusion/growth of BERD and creation of Innovation Capabilities A new intermediation form (between mutually adapted SU^ and VC^ & possibly the institutional framework) A critical mass of SU Other pre-emergence conditions
3	Strong Inn*(&Inn Pol.*) <->-Inn Finance* co-evolution, involving SU^ and VC^	'Quantitative Scaling Up' of SU^,VC^ i.e. Frequent, consistent and strong SU^-VC^ interactions (also mediated by enhanced presence of other agents)	Emergence of a VC market and industry Emergence of an ICT-oriented, high-tech, entrepreneurial cluster Quadruplicating of high-tech exports High proceeds from capital market operations (IPOs & M&As) Enhanced BERD/GDP & BERD/GERD; & of share of BERD financed by the business sector

6.2 Innovation policy and innovation finance implications

The existence of co-evolutionary processes enhances the likelihood that Innovation Policies broadly defined are linked over time. Thus, in our analysis, the successful emergence of a private Innovation Finance market (a VC market) during the 1990s highly depended on a prior successful *Innovation Policy*-Innovation** co-evolutionary process which started with the first Grants to BERD program in 1969 and ended which the highly successful VC-directed Yozma program 24 years later. This process was fueled both by the high responsiveness of business innovation to *direct* innovation policy measures and correspondingly, by the high responsiveness of innovation policy makers to the needs of the business sector – whether these needs required innovation policy measures like the above grants program or Yozma (i.e. whether directed to support company innovation or directed to support emergence of alternative private financial mechanisms).

In the Israeli case the early Grants to BERD program *contributed to create pre-conditions* for the successful implementation of the subsequent Yozma program 24 years later. This has numerous implications: notably, it is consistent with Lerner's 2009 assertion that building entrepreneurial systems requires policies and that these should be long run policies (Lerner 2009). It also implies that a major contribution (***the major*** contribution in the Israeli case) of the early Grants to BERD program was not direct but indirect, namely paving the way for the implementation of the highly successful Yozma program. Moreover, in these contexts, policies are not only the result of market/system failures but also contribute to generate such failures (Lipsey 2000; Avnimelech and Teubal, several papers).

The above has also strong possible implications for Venture Capital Policy. Rather than adopting a static perspective and advocating 'filling the early finance gap of existing SUs' (policies implemented during the 1990s in Europe, see various OECD reports) or promoting a VC market through tax benefits of various kinds and institutional changes and de-regulation, the approach that emerges from our analysis is Evolutionary (see Box 1 and last column of Box 6; and more extensively with adequate references in Rosiello *et al.* 2010, Avnimelech *et al.* 2009). One implication is that VC cannot be viewed as the main inducer of SU creation/development without prior, favorable pre-emergence conditions which include, among other things (the prior) existence of a critical mass of SUs (this would represent the so-called investment-ready opportunities or pent-up demand for VC services without which VC funds will not be established). Moreover, in many contexts there might be a *system failure* in achieving this critical mass of SUs, presumably a problem that plagued several large EU countries up to and including the 1990s. Our analysis and previous work with Avnimelech have extensively analyzed a number of other such conditions, including the existence of co-evolutionary processes such as those discussed in this paper.

6.3 Broader economic growth/development implications

1 *While strong growth of Business Innovation (I) is a requirement for successful Innovation and Structural Change (SC)-based Economic Growth, this paper suggests, based on the Israeli experience, that this may require sustainable Innovation Policy*-Innovation* and Innovation*-Innovation Finance* co-evolution* In Israel this occurred during a thirty year period which followed creation of a specialized agency in charge of Innovation Policy, namely the Office of the Chief Scientist at the Ministry of Industry and Trade. *Qualitative and non systemic co-evolutionary processes* predominating during the 1970s and 1980s; and *quantitative Innovation*-Innovation Finance* co-evolutionary process (dominated by SU-VC co-evolution)* dominating during the 1990s. These co-evolutionary processes which are necessarily linked with Innovation Policy and other favorable conditions as far as availability of skills and manpower are concerned, contributed to assure a sufficiently fast growth of business innovation after 1969 and the emergence of an Entrepreneurial, ICT-oriented, high-tech cluster (EHTC) during 1993–7/8. Alternatively, we may say that these co-evolutionary processes contributed to achieving Catching Up: a rapid diffusion of Innovation, increase in BERD and its share in GERD, while also contributing to create conditions for the emergence of the afore-mentioned entrepreneurial cluster.

2 *In the Israeli experience we see a strong initial focus on direct subsidy support of R&D in firms by the Government (starting in 1969).* This was characterized by continued adaptations of the mechanisms of support implemented, largely initiated as a result of policy learning and virtuous interaction with the private sector and its needs, as well as quantitative increases in the overall scope of support facilitated by the belief of the country's leaders at the time of the importance of developing 'science-based industry'. No less important was the search by policy makers (and facilitation given, sometimes through taking the initiative in the generation of specific programs like in the case of Yozma) of alternative, private sources of finance of business R&D. *In our opinion, no developing nor middle income country can afford not to consider the possibility of adopting subsidy or related schemes like conditional grants for the promotion of business innovation.*

3 *Assuring continued Innovation Policy*-Innovation* co-evolution requires both Policy Learning (already mentioned) and a favorable disposition to adapt policy to the changing 'needs' of the various segments of the business sector.* This will not always be the case, since it requires an entrepreneurial attitude on the part of leading policy makers which might not be there. Even more so is this a requirement for policy makers to actively be involved in searching for private sources of finance, since these could eventually substitute for those public sources whose disbursement they are in charge of undertaking. Israeli policy makers (first and foremost the Chief Scientists) were actively involved not only in adapting Government programs but also

were actively involved in searching for private, alternative sources of finance to satisfy the growing needs of firms.

4 *That experience suggests that policy makers in charge of Innovation Policy should make it their objective to ignite and sustain such co-evolutionary processes, as part of their overall remit and outlook of promoting innovation-based cumulative processes with positive feedback.* This is because, at least in certain cases and in the absence of broader information relating co-evolutionary process and business innovation, a necessary requirement for the eventual 'endogenization' and acceleration of business innovation seems to be precisely the presence of such co-evolutionary processes.

5 *The above is justified since such innovation-based cumulative processes are central for promoting innovation and structural change-based economic growth.* At an early stage (Israel prior to the 1970s, Korea prior to the 1980s approximately), the most relevant co-evolutionary process for igniting/sustaining such cumulative processes would be Innovation Policy*-Innovation* co-evolution (supposedly for some time during this early period, government funds would suffice to support the innovation needs of the business sector, so access to private sources of finance would be relatively less critical). Success in this goal would increase business sector R&D at a faster rate than otherwise. At some point this increase might outstrip the funding capacity of the Government, so the search for alternative, private sources of finance would then increasingly become critical. Igniting then a virtuous Innovation*(and Innovation Policy*)-Innovation Finance* co-evolution process would be key for continuation of the cumulative innovation-based growth process with positive feedback.

6 Whether this may or may not lead to the emergence of a Higher Level Organization such as an entrepreneurial high-tech cluster (and by extension, of one or more mid tech and even low tech cluster or set of clusters) may also depend on other factors. A major one will be a STE infrastructure generating high quality graduates in sufficient numbers to man the increasing number of R&D teams operating in the private sector (eventually explicit knowledge outputs will be required as well). In some countries the STE infrastructure grew fast rather exogenously from the processes mentioned above, due to Defense needs (many countries including Israel) and/or as in the case of Israel, cultural reasons. In general, exogenous factors (not directly relevant to the co-evolutionary processes considered) are critical for sparking and maintaining co-evolutionary process (see also Dutrenit *et al.* 2011): in the Israeli context these exogenous factors would include the aforementioned STE policies, the liberalization of trade, foreign exchange and capital markets of the mid 1980s and after, and 'global' exogenous factors as the ongoing ICT-related technological revolution including the global emergence of software as a separate industry during the early 1980s and of the Internet during the first half of the 1990s.

Notes

* This paper is based on Avnimelech and Teubal 2008a, 2009 and Teubal 2010.

1 See Avnimelech and Teubal 2006, 2008a,b,c and Teubal and Kuznetsov 2011.

2 This type of R&D could be termed 'regular' or 'classical' R&D to differentiate it from 'generic, cooperative' R&D, which is of a more infrastructural type (a generic, cooperative R&D program – the Magnet program – was implemented later in Phase 2). The latter's objective is to generate knowledge, capabilities and components rather than directly marketable outputs. The output of generic R&D would facilitate (or become inputs) to a subsequent 'regular' R&D activity.

3 Learning, including experience-based learning triggered by increased R&D in the Business Sector, was the main factor leading to enhanced R&D/Innovation capabilities.

4 The weak reported impact of Office of Chief Scientist support was probably also due to a 'technology bias' in the OCS R&D grants approval process.

5 VCs not only provided finance to SU but also 'added value', such as contributing to SU strategy, marketing, management, headhunting, access to complementary assets, reputation and networking.

6 Within an updated and adapted conceptual framework, Yozma is a *Framework Program* (see Section 5) whose implementation defines the initiation of Phase III. In this sense Yozma should be regarded as a Phase II–III interface program, with its design having been undertaken in Phase II while its implementation occurs in Phase III.

7 See Avnimelech and Teubal 2006, 2008a.

8 This fact might have been taken into account by Allen when he states 'The structural attractor therefore represents a reduced set of activities from all those possible in principle. It reflects the "discovery" of a subset of agents whose attributes and dimensions have properties that provide positive feedback". This would seem to imply that there are other subsets that could have been discovered which do not provide positive feedback. See below: the notion of 'dynamic economies of scale'.

9 Lerner mentions that this factor was more prevalent prior to the emergence of the modern SU and VC-based entrepreneurial system in the US during the 1980s and 1990s

10 This is consistent with a three-stage rather than a two-stage evolutionary process (Foster and Metcalfe, 2001, figure 1.1 and pp. 6–7).

11 VC learning was related to screening deal flow, due diligence, selection of investments, investing in startups, monitoring startups, providing value added, and in the exiting process. In the startup area, learning was related to structuring, raising capital, conducting effective R&D, marketing and alliance skills, penetrating global markets and other relevant skills. Collective learning stimulated entry of new market agents, and growth of existing market agents. Moreover, the learning process involved a significant component of *learning from foreign agents*: in the VC area particularly from the foreign LPs of Yozma Funds, and in the startup area particularly from foreign MNE.

12 Most of the material below was obtained from two interviews (January 1998 and May 2000) with Yigal Erlich the CEO of Yozma and one of the most important architects of the Program. Additional material was obtained from joint work with him at the UFISE and ESTER research programs during the years 2001–2006.

13 It should be noted that VC services at the time were 'non tradable'– that is, to a large extent they had to be provided within the cluster rather than imported. This changed to some extend during the late 1990s and early 2000s.

14 The incentives to the 'upside' also stimulated the entry of professional VC firms and managers.

15 This has been extensively analyzed in previous work, e.g., Avnimelech and Teubal 2006a, b.

16 In what follows we exclude in-house funding by firms and ignore tax concessions to BERD (of only fleeting existence in Israel during the 1980s) and other tax benefits of 'science-based' companies. Moreover, with the exception of the joint University-Business program Magnet, we largely ignore *indirect* support of innovation through support of the STE infrastructure since the bulk of their funding was exogenous to the processes being discussed. An exception was the restructuring of the Defense industries in the second half of the 1980s (see below), which by virtue of its strong direct impact on Civilian Innovation starting in the second half of the 1980s, will be considered as part of 'Innovation Policy'.

17 The star * indicates a broad definition of the relevant variables. The co-evolutionary processes will be termed (a) and (b) co-evolution respectively, or simply (a) and (b): (a) is the basic relationship and chronologically the first one to be identified in the Israeli scene. Both Innovation and Innovation Policy affected Innovation Finance and were affected by it, as shown in (b). When (a) is applied to Phase j (j=1,2,3) it will be expressed as (aj), and similarly with (b).

18 Elscint was the leading civilian, R&D- performing electronics firm during the 1970s (see Teubal 1986, Chapter 3).

19 See Section 2 of Pugatch *et al.* 2009.

20 An indication of the 'excess demand' for R&D funding at firms (relative to available resources from the OCS) was that at some point during the decade Elscint's budget requests alone represented almost half of all the OCS budget at the time.

21 This would seem to be an Angel or 'primitive' VC arrangement, since it doesn't correspond to the modern definition of Venture Capital (the so-called Lerner definition).

22 This view of the qualitative underpinning to the creation of a VC industry/market is to some extent implied in the analysis of Gompers and Lerner for the US despite the fact that the major thrust of their analysis lies in the 'institutional adaptations', particularly flexibilization of investment rules for Pension Funds (Gompers and Lerner 2001).

23 Note that the original objective of the Incubator Program was to provide a mechanism of creative and productive absorption of engineers and other high level personnel coming from the former Soviet Union. Note also that the budget of Magnet – which supported *generic, pre-competitive* R&D undertaken by consortia and Universities rather than 'regular' BERD which supported projects at individual firms – represented a net addition to Government disbursements rather than a redistribution of existing resources.

24 Correspondingly it also signalled the existence in Israel at the time of adequate Investment-Ready opportunities or Deal-Flow for VC. For the notion of Investment-Ready opportunities see Mason and Harrison 2003.

25 For example, frequently entrepreneurs who benefited from VC investments subsequently become consultants to the VC or even partners or founders of new VCs (VC interviews during the years 1999–2001, and Gompers *et al.* 2005).

References

Allen, P. (2004), "The Complexity of structure, strategy and decision making", Chapter 5 in J. S. Metcalfe and J. Foster (eds) (2004) *Evolution and Economic Complexity,* Edward Elgar Publishing Limited: Cheltenham, Glos GL50 1 UA, UK

Avnimelech, G. and M. Teubal (2004), "Venture Capital–Startup Co-Evolution and the Emergence and Development of Israel's New High Tech Cluster", *Economics of Innovation and New Technology,* 13(1), pp. 33–60.

Avnimelech, G. and M. Teubal (2006a), "Creating VC industries which co-evolve with High-tech: Insights from an Extended Industry Life Cycle (ILC) perspective to the Israeli Experience", *Research Policy,* Vol. 35(10), pp. 1477–1498. Special Issue on

"Mapping the links between Universities, the Economy and Government: Combining Scientometrics and Evolutionary Economics", L. Leydesdorff and M. Meyer (Eds).

Avnimelech, G. and M. Teubal (2006b), "The Emergence of Israel's Venture Capital Industry: How Policy can Influence Cluster Dynamics", in P. Braunerhjelm and M. Feldman, *Cluster Genesis: Technology-Based Industrial Development*, Oxford University Press, Oxford.

Avnimelech, G. and M. Teubal (2008a), "From Direct Support of Business Sector R&D/Innovation to Targeting Venture Capital/Private Equity: A Catching Up Innovation and Technology Policy Life Cycle Perspective", *Economics of Innovation and New Technology*, 17(1 & 2), pp. 153–172, January-March

Avnimelech, G. and M. Teubal (2008b): "Evolutionary Targeting", *Journal of Evolutionary Economics*, 18(2), 151–166

Avnimelech, G. and M. Teubal (2009), "The Co-evolution of ICT, VC and Policy in Israel during the 1990s." in D. Fornhal, S. Henn and M.-P. Menzel (eds) *Emerging Clusters*, Elgar 2010.

Avnimelech, G., A. Rosiello, and M. Teubal (2010), "Evolutionary Interpretation of Venture Capital Policy in Israel, Germany, UK and Scotland", *Science and Public Policy* 37(2), March

Bresnahan, T. and A. Gambardella (Eds) (2004), *Building High Tech Clusters: Silicon Valley and Beyond*, Cambridge University Press, Cambridge.

Bresnahan, T., A. Gambardella and A. Saxenian (2001), "Old Economy Inputs for New Economy Outputs: Cluster Formation in the New Silicon Valleys", *Industrial Corporate Change,* 10(4), 835–860.

Breznitz, D. (2007), "Industrial R&D as a national policy: Horizontal Technology Policies and industry-state co-evolution in the growth of the Israeli software industry", *Research Policy*, 30.

Breznitz, D. (2008), *Innovation and the State, Political Choice and Strategies for Growth in Israel, Taiwan and Ireland*, Yale University Press, New Haven and London

CBS (2008). *Central Bureau of Statistics*, Israel, Available at: www.cbs.gov.il

Dutrenit, G., M. Puchet and M. Teubal (2011), "Building bridges between co-evolutionary approaches to Science, Technology and Innovation and Development Economics: an interpretative model", in *Innovation and Development*, Vol. 1, Iss.1 51–74.

Foster, J. and J.S. Metcalfe, *Frontiers of Evolutionary Economics: Competition, Self-Organization and Innovation Policy*, Edward Elgar: Cheltenham, UK and Northhampton, MA, USA

Gompers, P. and J. Lerner (1999), *The Venture Capital Cycle*, Cambridge, Massachusetts and London, England: The MIT Press.

Gompers, P. and J. Lerner (2001), "The Venture Capital Revolution", *Journal of Economic Perspectives*, 15(2), pp. 145–168

Gompers, P., J. Lerner and D. Scharfstein (2005), "Entrepreneurial Spawning: Public Corporations and the Genesis of New Ventures, 1986–1999", Journal of Finance 60 (2005): 577–614.

IAEI (2008). *Israel Association of Electronics and Information Industries*. (www.iaei.org.il), Israel.

IVC (2008). *Israel Venture Capital Research Center Database* (www.ivc-online.co.il).

Lerner, J, (2009), *The Boulevard of Broken Dreams: Why Public Efforts to Boost Entrepreneurship and Venture Capital have Failed – and what to do about it*, Princeton University Press, Princeton, New Jersey; especially Chapters 6 and 7

Lipsey, R., Mason, C. and R. Harrison (2003), Closing the Regional Equity Gap? A

Critique of the Department of Trade and Industry's Regional Venture Capital Funds Initiative, *Regional Studies*, 37(8), 855–868

Murmann, J.P. (2002), "The Coevolution of Industries and National Institutions: Theory and Evidence", working paper, FSIV02.14, Social Science Research Center Berlin.

Murmann, J.P. (2003), "The Coevolution of Industries and Academic Disciplines", working paper, WP03–1, Kellogg School of Management, Northwestern University.

Nelson, R. (1994), "The Co-evolution of Technology, Industrial Structure and Supporting Institutions", Industrial and Corporate Change

Nelson, R. (2001), "The coevolution of technology and institutions as the driver of economic growth", Chapter 2 in J. Foster and J.S. Metcalfe, *Frontiers of Evolutionary Economics: Competition, Self-Organization and Innovation Policy,* Edward Elgar: Cheltenham, UK and Northhampton, MA, USA

OCS (2008), *Office of Chief Science*, Ministry of Industry, Trade and Labor (www.moit.gov.il), Israel

OECD, (2004), Venture capital: trends and policy recommendations. DSTI/DOC(2004), *STI Working Papers*, Paris.

Rosiello, A., G. Avnimelech and M. Teubal (2009), "Towards the Framing of Venture Capital Policies: A Systems Evolutionary Perspective with Particular Reference to the UK/kScotland and Israeli Experiences", *Journal of Evolutionary Economics* 2010, Special Issue, forthcoming. Proceedings of the International Schumpeter Society Meeting, Rio de Janeiro July 2008

Saxenian, A. (1994), *Regional Development: Silicon Valley and Route 128*, Harvard University Press.

Teubal, M. (1983), "Neutrality in Science Policy: The Promotion of Sophisticated Industrial Technology in Israel", Minerva 21: 172–179; chapter in M. Teubal 1987: Innovation Performance, Learning and Government Policy, Selected Essays, The University of Wisconsin Press.

Teubal, M. (1987), Innovation Performance, Learning and Government Policy, Selected Essays, The University of Wisconsin Press.

Teubal, M. (1997), "A Catalytic and Evolutionary Approach to Horizontal Technological Policies", *Research Policy*, 25, pp. 1161–1188.

Teubal, M. (1999), "An R&D Strategy for Israel", Economic Quarterly (Hebrew), 46, November, pp. 359–383

Teubal, M. (2008), "Direct Promotion of Commercial Innovation (CI) in Least Developed Countries: A Systems Evolutionary (S/E) Perspective" background paper for UNCTAD's Least Developed Country's 2009 report

Teubal, M. (2010), "Notes on Innovation-Innovation Finance Co-evolution". Presented at Hohenheim University, Stuttgart, October

9 Venture capital in Canada

Jorge Niosi

Introduction

Venture capital (VC) is a financial industry that plays a key role in the renewal of the business sector in all developed and emerging countries. In most countries, venture capital supports high-technology start-ups (Avnimelech, Schwartz and Teubal, 2008; Florida and Kenney, 1988, Gompers and Lerner, 1999; Niosi, 2000). Thus, a strong VC industry is a major factor behind the rise of new industries and the development of new technologies. It contributes not only to the financial and managerial support of new industries but also to the development of linkages between research universities, large corporations, business angels, and clients or users of the new products.

The United States has led the world in the development of this new industry and in 2008 still represented 49% of OECD venture capital investment (OECD, 2009). US financial investment in venture capital has been emulated by a series of other OECD countries including Canada, France, Germany, and Great Britain, among others. The list of successful followers in this industry, however, is much shorter than the list of failures (Lerner, 2009).

Canada, together with Israel and Britain, has been one of the few accomplished imitators of the US venture capital industry, even if the present financial crisis has affected the VC industry in all three countries. This paper presents the key definitions (section 1), a theoretical discussion on venture capital follows (section 2), then analyzes the Canadian venture capital industry from its origins to the present day (section 3). The international context follows (section 4), and a discussion and policy implications.

1 What is venture capital?

The definitions of venture capital vary widely from one country to the next, not only in theoretical terms but also, more important, in legislation and business practices. The OECD defines venture capital as "capital provided to young unquoted firms with high growth potential." (OECD, 2011: 102) In Canada, venture capital is capital usually invested in technology companies not guaranteed by the assets of the investee. The frontiers between venture capital and angel capital are sometimes not very clear, especially when angels organize themselves in associations and copy the investment patterns of successful venture capital firms. Gompers and Lerner (1999: 343) defines an angel as "a wealthy individual who invests in entrepreneurial firms. Although angels perform many of the same functions as venture capitalists, they invest their own capital rather than that of institutional and other individual investors." Angels are often supposed to fill the gap between the investments made by the original entrepreneurs and their close relatives (fools, family and friends) and venture capital that comes later. In Canada, in 2008, angels are supposed to have invested some $3 billion in new ventures against $1.6 billion by VC investors, according to Dow Jones. Like VC, angels request equity in exchange for their investments, which usually occur in the form of seed capital, during the very early stages of a new technology-based firm.

Venture capital is made of "independently managed, dedicated pools of capital that focus on equity or equity linked investments in privately held, high-growth companies" (ibid.: 349). VC firms are most often limited partnerships that use their own funds plus the funds they collect from wealthy individuals, financial institutions and other organizations. Angels tend to remain passive investors, while most the managers of venture capital firms are actively involved in the management of the start-up company.

Private equity markets (PEM) are those that serve usually mature (meaning generating cash flow) companies not publicly traded in the stock exchanges. Private equity markets, like venture capital markets, are equity ones. The difference is the age of the investee companies: VC firms invest in new companies, while PEM serve established private firms. Yet, in many countries, figures of private equity are presented in conjunction with venture capital data, contributing to the confusion of both types of investments. PE firms however tend to abhor high-tech sectors and invest in traditional commercial, financial and manufacturing businesses. According to McKinsey, in 2008 there were some 190 active PE funds managers in Canada, with a total capital under management of C$84.7 billion, and they had disbursed 11.9 billion that year (McKinsey & Co., 2009).

Captive VC firms existed since the 1980s mainly in the United States (McNally, 1997). Johnson and Johnson Development Corporation (organized in 1973 is one of the oldest. Intel Capital (founded in 1998) is a more recent appropriate example. After the year 2000 and the crash of the stock markets, a large number of high-technology users – including pharmaceutical and information technology corporations – launched captive VC firms in order to invest in promising new technologies being developed by new start-up. Takeda Research Investments (established in 2002) and Pfizer Venture Capital Investments, set up in 2004, are cases in point. These captive funds added themselves to those already in existence, previously established by banks, industrial and insurance companies. Corporate venture capital is thus different from limited partnerships firms, and often invests in one or two specific sectors, linked with the particular interests of the corporate holding. Sometimes authors make the difference between "internal corporate VC" when a large company invests in a spin-off firms using technology generated in the mother company, and "external VC" when the same large company invests in an entirely independent new technology-based firm.

2 Theories: failures, failures, failures

Venture capital industries have emerged to address a most important market failure: established commercial and investment banks have developed business routines to evaluate physical assets such as real estate, commodities, machinery or stocks, but not to evaluate intangible assets such as chemical formulas, software algorithms, genetically modified organisms or R&D projects for new drugs. In some cases, they have developed their expertise, real or imaginary, to evaluate derivative products or sovereign debt. In a number of cases (including in Canada) established private and public banks have created independently managed subsidiaries whose mission is to invest in venture capital.

Another key dimension of the failure problem is information asymmetries: entrepreneurs may overstate the value of their technology without investors noticing it, or they may be reluctant to entirely disclose it to third parties because they risk losing it. Also, VC investors could use or reveal to third parties the information they receive from entrepreneurs.

In order to reduce these information asymmetries without entirely eliminating them, VC firms tend to stage their investments in different rounds of increasing investment (seed capital, start-up capital, first round for market test and initial manufacturing, working capital, first expansion, second expansion, and IPO). If the project proceeds in the right direction, the value of the investment and the underlying technology should increase, and the risk should decrease.

But venture capital is not only the response to ***market*** failures. It is also the response to ***system or institutional*** failures. It has often occurred that venture capital develops because laws and regulations about commercial or investment banking do not allow these financial organizations to invest in risky businesses. Sometimes, some kind of policy does not allow pension funds to invest part of their assets in venture capital. Pension funds were at the origin of American (and Canadian) venture capital. In some countries, pension funds are not permitted to invest in venture capital thus closing this avenue to the rise of a VC industry.

Another type of government failure may occur when governments invest directly in venture capital, through public VC corporations. The failures here are of several different natures. First, government VC tends to be a passive investor. yet private VC, as was emphasized before, nurtures the development of new firms, not only by funding them, but also by helping the managers of the start-up to put forward appropriate financial, marketing and production strategies. They also link the new form with other corporations (financiers, clients, suppliers, users). They bring credibility to the new company, and carefully monitor the start-up.

Public VC investors mostly provide funds to the new firm. They are usually less involved in the management of the start-up. Their selection of projects may also be affected by political considerations (geographical representation, ethnic considerations or political links). The result is that most often public VC companies exhibit negative returns and are closed up after an evaluation shows their poor bottom lines.

Even if public intervention is often associated with VC failures (Lerner, 2009), governments have played a major role in the emergence of a VC industry most remarkably in the United States. They have done so through several mechanisms. First, by supporting the growth of advanced technologies in universities and public laboratories, a condition for the demand for venture capital. Second, by allowing the investment of pension funds in venture capital firms, and by reducing taxes through the New Market Tax Credit incentive (federal level) as well as by similar incentives at the state level. Third, by supporting the growth of these technology-based start-ups through non-reimbursable subsidies to R&D through such US federal programs as SBIR (Small Business Innovation Research program) STTR (Small Business Technology Transfer program) and state programs such as the Ben Franklin Technology partnership (Pennsylvania) and TEDCO (Maryland) providing non-reimbursable funds to technology firms and thus helping start-ups during their most risky period of growth.

Somewhere inbetween public and private, the Yozma program in Israel appeared as a best practice in government promotion of venture capital. Yozma was launched in 1993 by the Israeli government to fund private-sector VC firms, thus increasing the available funds for VC investment. Yozma was designed as a fund of funds. Instead of trusting public managers to select investee companies, the Israeli government endowed private experienced managers with the administration of the Yozma funds. Yozma I (1993–1998) invested $100 million in 8 different private funds, thus nurturing the emergence of the Israeli VC industry (Avnimelech, 2009). Yozma was a total success, probably not only because the public program was well designed and implemented, but also because a large proportion of the funds were linked to foreign well known VC firms based in North America and Western Europe. Yozma raised $263 million (of which $100 million was from the Israeli government) invested in 164 start-up firms. The exit rate was very high: over 50% of the start-ups went to a particular exit, and only 25% closed.

3 The Canadian VC industry

The federal banking law of 1871 created a system of large federal banks with a clear orientation towards commercial lending. Provinces were not allowed to give charts to commercial banks, while in the United States the law authorised states to incorporate banks. Investment banking and insurance remained separate businesses. Thus the Canadian banking system remained more solid than the US one, but also more conservative than the US financial system (Neufeld, 1972). By the end of the nineteenth century and beginnings of the twentieth century other financial intermediaries appeared, particularly the investment banks, helping governments, corporations and individuals in raising finance through underwriting securities. They also intervene in mergers and acquisitions, but in contrast to commercial banks, these investment banks do not take deposits or extend loans. Yet, investment banks also acquire securities of private firms, acting in the private equity markets, at that time not quite different from VC markets. Canada's large commercial banks have established investment-banking subsidiaries. Again, in contrast to the US the funds of the commercial banks and their investment banking divisions remain separate. As such, the Canadian banking system is synonymous with conservative management and little interest in elaborate derivative products or risky government debt.

In such an environment the VC industry started slowly through the investment of American and British VC subsidiaries in Canada, then increased by the addition of a Canadian financial innovation; the Labour-sponsored venture capital corporations (LSVCC) in 1984, backed by federal and provincial tax relief for individuals. Individuals could buy shares of these LSVCC for their retirement plans and could deduct a large percentage of this investment from their tax payments. In the early 2000s, LSVCC represented up to 40% of the entire Canadian VC industry.

By 2007, Canada ranked 14th among OECD countries by the ratio of venture capital as a percentage of GDP. The list was headed by Denmark with 0.359%, the United States was in 5th place with 0.225% and Canada trailed with 0.134% (OECD, 2009b). With 0.4% Canada was still in 5th place in 2001, after the United Kingdom, Sweden, the United States and the Netherlands.

Like its United States counterpart, Canadian VC industry experienced a fast increase in the funds it managed, in its investments, and the number of companies supported, and a strong decline after year 2000 followed by September 11, 2001 and the present day financial crises (see Figure 9.1).

In Canada, the demand for VC funds has been strong. It comes both from the academic and private sectors. Canada has one of the most productive university systems in terms of publication, patenting and spin-off foundation. Government support for academic research is one of the highest in the world: by 2007, Canada's higher education expenditure on R&D (HERD) represented 0.68% of GDP, against 0.43% in Japan, 0.41% in Germany, 0.40% in France, and 0.36% in the USA (OECD, 2009). Only the Nordic countries have a higher % of HERD on GDP. Also, over 600 foreign companies have R&D labs in Canada, out of over 20000 firms conducting R&D in the country every year. There is a strong demand for VC from private sector spin-offs. According to Statistics Canada, over 1200 academic spin-offs were created up to December 2008, mostly in Montreal, Toronto and Vancouver. Spin-offs from large private R&D intensive companies are numerous; they include those of companies such Bell Canada, Bombardier, CAE, HQ and Ericsson Canada in Montreal, IBM Canada and Magna in Toronto, Cognos and Nortel in Ottawa, or RIM in Waterloo (ON), hosting thousands of researchers each. At the last count, there are 668 biotechnology companies in Canada of which over 90 publicly quoted (mostly academic spin-offs) and over 6000 software companies of which some 100 quoted (mostly private sector spin-offs); 44 Canadian technology-based companies are quoted in

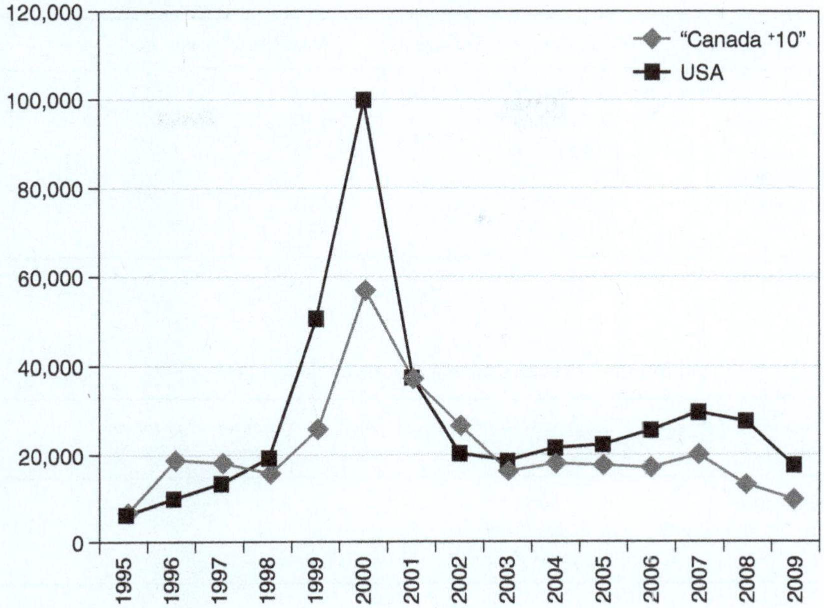

Figure 9.1 US and Canadian VC investments (billions of current $) (sources: Canadian Venture Capital Association (CVCA) and National Venture Capital Association (USA)).

NASDAQ (but just 1 from Brazil, 1 French, 2 German, 3 Indian, 1 Italian, 1 from Sweden, and 5 British companies). Only Israel, with 58 companies, is ahead of Canada in the number of companies quoted in NASDAQ.

Tables 9.1, 9.2 and 9.3 bring recent information about VC activities in Canada, by region and technology, as well as on new versus follow-on investment. These tables show that two provinces (Québec and Ontario, which are also regions) concentrate over two-thirds of VC investment in Canada, followed by British Columbia. Two other regions, comprising seven provinces, only represent 15% of total VC activity. Venture capital is enormously concentrated in the

Table 9.1 Regional distribution of venture capital in Canada, 2009

Region/province	*Investment in C$M (and %)*	*Number of companies financed*
Quebec	431 (43%)	159
Ontario	288 (29%)	87
British Columbia	141 (14%)	39
Prairies	82 (8%)	29
Atlantic provinces	68 (7%)	17
Total Canada	1,010 (100%)	331

Source: CVCA.

Table 9.2 Sector distribution of venture capital in Canada, 2009

Sector	*Investment in C$M (and %)*	*Number of companies financed*
Life sciences	215 (20%)	57
ICT	489 (49%)	129
Other high technologies	115 (12%)	29
Traditional	190 (19%)	116
Total	1,010 (100%)	331

Source: CVCA.

Table 9.3 New and follow-on venture capital investment in Canada, 2009

New versus follow-on investment	*Investment in C$M (and %)*	*Number of companies financed*
New investment	346 (34%)	146
Follow-on investment	663 (66%)	191
Total	1,010 (100%)	331

Source: CVCA.

largest cities (Toronto, Montreal, Vancouver and Ottawa) and in the largest and most densely populated provinces. Also, high-tech activities represent over 80% of Canadian VC investment, with life sciences and information and communication technologies keeping the lion's share. Finally, follow-on investments represent well over 50% and close to two-thirds of all VC investment in Canada; this is part of a trend of VC moving away from high-risk activities and investments and into more conservative, less risky investments.

The geography and sector distribution of Canadian VC industry follows basic laws of supply and demand: where demand is resilient and strong, supply will pop up. In Canada, demand is strong in the three most populous provinces, where most of the largest research universities are located: Ontario, Quebec and BC. Thus the Prairie Provinces and the Atlantic Provinces suffer from both a lack of demand and an investment gap: few VC firms are located there and few foreign funds support the scarce local high-tech firms. This fact, by the way, confirms the demand and supply approach to VC: a country (or region) can only grow a VC industry if it has a vibrant higher education set of institutions conducting research.

Figures 9.2 and 9.3 show that the number of initial public offerings (IPOs) is now reduced to almost nothing, and the number of mergers and acquisitions (M&A) has grown to fill the gap, and then stabilized between 20 and 25 per year. Also, the average value of these IPOs has declined, and the value of the M&A offerings has grown from C$40 to C$120 million between 2003 and 2009.

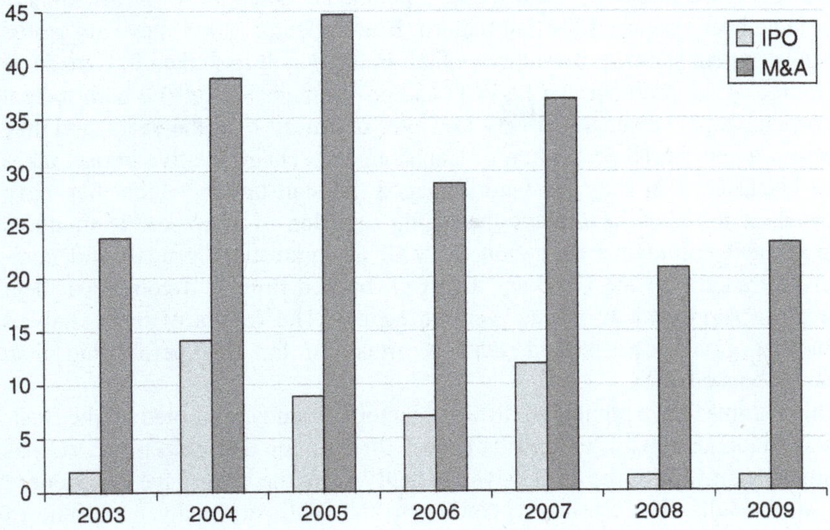

Figure 9.2 Number of Canada's VC-Backed IPOs and M&A, 2003–2009 (source: Canadian Venture Capital Association and Thomson Reuters).

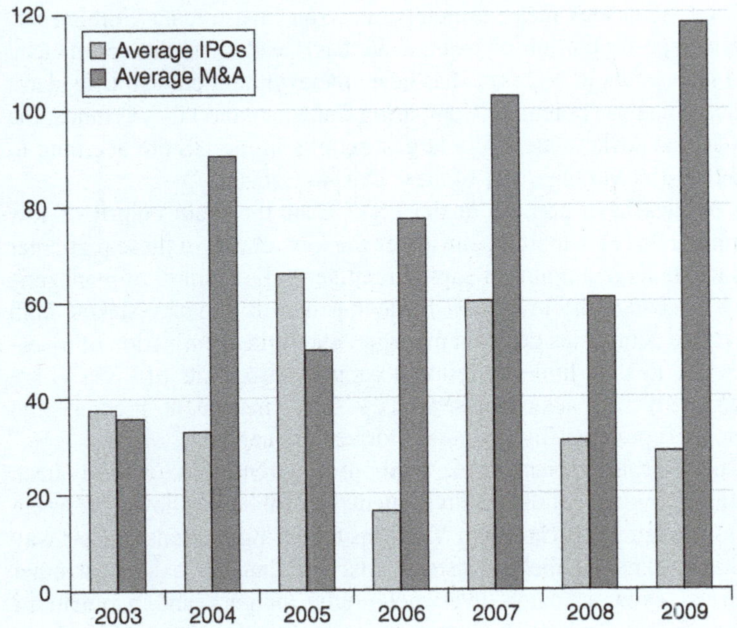

Figure 9.3 Average disclosed offerings in VC-Backed IPOs and M&A, 2003–2009, C$M (source: Canadian Venture Capital Association and Thomson Reuters).

However, Canada's VC industry lags behind its potential. Several explanations have been proposed for that pattern. Brander *et al.* (2008) have suggested that government sponsored venture capital, in the way it has taken in Canada, is the source of the problem: the LSVCC and government VC (GVC) both federal and provincial, have generated very low rates of returns over the years, and they represent a substantial proportion of Canadian VC. Their negative impact takes place because either they have mismanaged the start-ups in which they have invested, or that they have made the wrong selection of investees. The authors have studied employment creation, as well as innovation, growth and competitive strengths of the GVC-and LSVCC- backed firms and compared them with those supported by private venture capital. The results of their analysis show that government-backed start-up firms are far less performing than privately-backed ones.

Other studies have pointed to different factors. A study promoted by the Business Development Bank of Canada (BDC, 2011) finds that not only GVC has obtained negative returns; most privately held VC partnerships have experienced the same results. The reasons, according to the BDC, are many. Only one of them refers to the role of government-backed venture capital. Other reasons prevail. There is a shortage of serial entrepreneurs in Canada. General partners in VC firms are subscale and lack strong management capabilities and experience compared to their US counterparts. The angel network is not well developed. There is a marked overinvestment in seed stages without enough investment on follow-on phases, leading to dilution and lack of continuity. Undercapitalised syndicates make difficult follow-on investments. Foreign VC firms capture a large proportion of exits. Exits have lead to mediocre financial returns. R&D investment of VC firms has been made without enough consideration of commercialization potential. Comparing Canadian and US VC funds, the BDC concludes that scale matters: the largest returns in the US are accruing to funds in the $200–300 M range; few of these exist in Canada.

Still other studies have pointed to the LSVCC as the main culprit of low returns (Cumming, 2011). Cumming attributes the low returns of these particular mutual funds to the large amount of capital relative to the number of managers, lack of ability to change the investees' behaviour due to minority stakes, high number of investee companies per fund manager, and little examination of investee candidates due to time limits to reinvest capital. Also, exits of LSVCC are most often company buybacks and secondary sales, instead of mergers and acquisitions or IPOs preferred by private-sector venture capital.

In spite of these negative elements, several positive trends have recently been observed. In the last years, follow-on investments, as Table 9.3 shows, represent over 50% of all investments. Canadian VC firms are trying to manage their way towards profitable exits for their investees. Also, in Canada as in most other OECD countries, exits are now overwhelmingly mergers and acquisitions (M&A) instead of IPOs. These M&A are more profitable than IPOs. Also, Canadian VC firms increasingly invest abroad and foreign VC invests in Canada, thus concentrating their investments on the best potential investees. More important, since 2007, governments have learned and they are increasingly investing in "funds of funds" in a way that recalls the Yozma funds and the French "SME and Innovation program (Giorgino, 2006). Canadian federal and provincial governments are putting public funds in the hands of experienced VC private investors. Table 9.4 shows some recent cases.

Other changes include a different geographical destination of funds. Historically, Ontario as the largest province was the main location of VC investments. Ontario hosts two ICT clusters (Toronto and Ottawa) and one biotechnology cluster (in Toronto). The Quebec province hosts two biotechnology clusters (Montreal and Quebec City) and one ICT cluster in Montreal. This regional change may be related to the presence of the large Teralys program, but it may be too early to say whether this is a permanent trend or a short-term blip.

All these national changes occur within major changes in the international context. First, we witness a relative decline of North American VC investment following the dot.com crisis, September 11, and the current financial crisis. We also see a relative rise of China, India and Western European VC funds, in percentage of global funds, within a major crisis of venture capital investments affecting all OECD countries and Israel.

Second, VC becomes more internationalized, with international syndicates merging and investments crossing borders not only within regions, such as North America or Western Europe, but also continents. Five major source countries are emerging: the United States, the United Kingdom, Germany, Canada and France, in that order. The major recipient countries are the United Kingdom, China, Canada, India, France and Sweden.

Third, venture capital has become more selective, reserving investment for existing portfolios, faster returns and follow-on disbursements.

Conclusion

Venture capital, in Canada as elsewhere, has been shown to be very dependent upon economic cycles. In the present downturn period, venture capital is stalling. Yet, Canada suffers from problems of its own, most probably from small

Table 9.4 New Canadian funds of funds

Name of fund	*Month and year established*	*Amount of fund*
Ontario VC Fund	November 2007	C$90 million
Alberta Enterprise Corp.	April 2008	C$200 million
BC Renaissance Fund	April 2008	C$80 million
Teralys Fund (Quebec)	April 2009	C$700 million
Ontario Emerging Technology Fund	October 2009	C$250 million
BDC	November 2009	C$90+C$75 million
Total	2007–2009	C$1,485 billion

Source: author's own compilation.

regional markets, excessive role of LSVCC and government venture capital, reduced entrepreneurship and an underdeveloped network of angel investors.

Under the new context of increasing demand from China and India, and a requirement of higher returns and faster exits, some Canadian funds have started to internationalize. Also:

> **Follow-on capital increases:** Both Canadian and US VC firms investing in Canada are turning towards follow-on rounds in order to increase the chances that their portfolios become profitable
> **Governments learn:** they have curtailed their VC direct investment in new technology-based firms and instead launched VC "funds of funds" programs, similar to the Israeli Yozma program, which is widely considered as a best practice. A major difference with Yozma is that the Israeli fund aimed at creating a brand-new VC industry in Israel, while the Canadian government aim is to revive an ailing VC industry. Canadian new course shows a strong capacity to learn from government policymakers.
> **A modest revival of the North American VC industry:** There are preliminary indications that the VC capital market is reviving in North America due to the rebound of the stock markets in 2010, and continuous efforts by governments to revamp the VC industry

In 2010, more new IPOs are appearing and others are being prepared.

Canadian stock exchanges are attracting foreign VC firms to float their new technology-based firms in the Toronto Stock Exchange and the TSX Venture Exchange.

Yet, time will tell whether these changes are the result of government efforts, or are short-term trends without a future due to the persistent financial crisis affecting Western Europe and the United States.

Bibliography

Avnimelech, G. (2009): "VC Policy: Yozma Program 15 Years Perspective", a presentation to the 2009 DRUID Summer Conference, Denmark.

Avnimelech, G., D. Schwartz and M. Teubal (2008): "Venture Capital Emergence and Start Up-Intensive High-Tech Cluster Development: Evidence from Israel", in C. Karlsson (ed.): *Handbook of Research on Innovation and Clusters*, Cheltenham, Elgar, pp. 124–146.

Brander, J. A., E. J. Egan and T. F. Hellman (2008): *Government Sponsored versus Private Venture Capital: The Canadian Evidence*, Cambridge, MA, NBER Working paper 14029.

Business Development Bank of Canada (2011): *Venture Capital Industry Review*, Ottawa, BDC.

Cumming, D. (2011): "Review Essay: Public Policy and the Creation of Active Venture capital markets", *Venture Capital: An International Journal of Entrepreneurial Finance*, (Forthcoming).

Florida, R. and M. Kenney (1988): "Venture capital-financed innovation and technological change in the USA", *Research Policy*, 17 (3): 119–137.

Giorgino, M. (2006): "Public models to support venture capital market", *International Journal of Entrepreneurship and Innovation Management*, 6 (4/5): 444–461.

Gompers, P. and J. Lerner (1999): *The Venture Capital Cycle*, Cambridge, MA, MIT Press.

Lerner, J. (2009): *Boulevard of Broken Dreams, Why Public Efforts to Boost Entrepreneurship and Venture Capital has Failed and what to do about it*, Princeton, Princeton University Press.

McKinsey & Co (2009): *Private Equity Canada 2008, the Industry at an Inflection Point*, Toronto, McKinsey and Co.

McNally, K. (1997): *Corporate Venture Capital Bridging the Equity Gap in the Small Business Sector*, London, Routledge.

Neufeld, E. (1972): *The Financial System of Canada, its Growth and Development*, Toronto, Macmillan.

Niosi, J. (2000): *Canada's National System of Innovation*, Montreal, McGill-Queen's University Press.

OECD (2009): Science, Technology and Industry Scoreboard 2009, Paris.

OECD (2009b): OECD Economic Surveys: Italy 2009, Paris.

OECD (2011): Entrepreneurship at a Glance 2011, Paris.

Index

Page numbers in *italics* denote tables, those in **bold** denote figures.